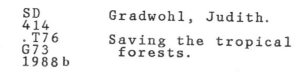

Saving the Tropical Forests

FUNDS TO SUPPORT THIS PUBLICATION OF

Saving the Tropical Forests

WERE PROVIDED BY

THE JOHN D. AND CATHERINE T.

MACARTHUR FOUNDATION

Saving the Tropical Forests

by JUDITH GRADWOHL
and RUSSELL GREENBERG

Preface by Michael Robinson
Illustrated by Lois Sloan

ISLAND PRESS

Washington, D.C. □ *Covelo, California*

ABOUT ISLAND PRESS

Island Press, a nonprofit organization, publishes, markets, and distributes the most advanced thinking on the conservation of our natural resources—books about soil, land, water, forests, wildlife, and hazardous and toxic wastes. These books are practical tools used by public officials, business and industry leaders, natural resource managers, and concerned citizens working to solve both local and global resource problems.

Founded in 1978, Island Press reorganized in 1984 to meet the increasing demand for substantive books on all resource-related issues. Island Press publishes and distributes under its own imprint and offers these services to other nonprofit organizations.

Funding to support Island Press is provided by The Mary Reynolds Babcock Foundation, The Ford Foundation, The George Gund Foundation, The William and Flora Hewlett Foundation, The Joyce Foundation, The J. M. Kaplan Fund, The John D. and Catherine T. MacArthur Foundation, The Andrew W. Mellon Foundation, Northwest Area Foundation, The Jessie Smith Noyes Foundation, The J. N. Pew, Jr. Charitable Trust, The Rockefeller Brothers Fund, and The Tides Foundation.

For additional information about Island Press publishing services and a catalog of current and forthcoming titles, contact Island Press, P.O. Box 7, Covelo, California 95428.

First published 1988 by Earthscan Publications Limited, London. This Island Press edition is printed by special arrangement with Earthscan Publications Limited.

Cover design, Studio Grafik. Cover photograph, Steve Brosnahan.

Library of Congress Cataloging-in-Publication Data

Gradwohl, Judith.
  Saving the tropical forests.

  Bibliography: p.
  Includes index.
  1. Forest conservation—Tropics. 2. Deforestation—Control—Tropics. 3. Forest reserves—Tropics. 4. Forest management—Tropics. 5. Reforestation—Tropics. 6. Rain forest ecology. 7. Sustainable agriculture—Tropics. I. Greenberg, Russell. II. Title.
SD414.T76G73 1988b   333.75'16'0913   88-31267
ISBN 0-933280-81-5

Manufactured in the United States of America
10 9 8 7 6 5 4 3 2

CONTENTS

Preface: Beyond Destruction, Success (by Michael H. Robinson) 11

Foreword 17

Acknowledgements 21

Part I: The Specter of Deforestation 23

Part II: The Case Studies 53

CHAPTER 1: FOREST RESERVES 57

The Sian Ka'an Biosphere Reserve: Conservation of Forest
and Sea, Mexico [1] 70

The Community Baboon Reserve: An Approach to the
Conservation of Private Lands, Belize [2] 72

Land Titling and Forest Protection around the
Gandoca/Manzanillo Wildlife Refuge, Costa Rica [3] 76

La Amistad Biosphere Reserve, Costa Rica [4] 78

The Kuna Yala Biosphere "Comarca": An Indigenous
Application of the Conservation Concept, Panama [5] 81

A Bi-National Approach to the Protection of Indian
Lands, Colombia and Ecuador [6] 83

The Cuyabeno Wildlife Production Reserve, Ecuador [7] 85

Manu Biosphere Reserve, Peru [8] 88

Protection and Development in and about Khao Yai Park,
Thailand [9] 91

Protecting Wildlife and Watersheds at Dumoga Bone,
Indonesia [10] 94

A Community-Managed Buffer Zone for a Nature Reserve,
Indonesia [11]   96

Korup National Park, Cameroon [12]   99

CHAPTER 2: SUSTAINABLE AGRICULTURE   102

Lessons from Mayan Agriculture, Central America [13]   110

An Intensive Agricultural System for Forests with
Karsted Limestone Areas, Mexico [14]   113

Converting from Beef to Dairy Cattle, Costa Rica [15]   116

Iguana Ranching: A Model for Reforestation, Panama [16]   118

Resource Management by the Kayapó, Brazil [17]   122

Japanese Farming in the Amazon Basin, Brazil [18]   125

Long-term Cultivation of Swidden-Fallows by
Bora Indians, Peru [19]   127

Market-oriented Agroforestry in the Amazon, Peru [20]   129

Javanese Home Gardens, Indonesia [21]   132

An Extension Service for Shifting Agriculturalists,
New Guinea [22]   134

CHAPTER 3: NATURAL FOREST MANAGEMENT   138

A Sustainable Silvicultural System for Forests,
Suriname [23]   144

Harvesting the Flood Plain Forests, Brazil [24]   147

Extractive Reserves: A Sustainable Development
Alternative for Amazonia, Brazil [25]   149

Natural Forest Regeneration and Paper Production,
Colombia [26]   153

Public and Private Cooperation in Protecting and Managing
a Tropical Watershed, Colombia [27]   155

Sustained-Yield Management of Natural Forests in the
Palcazu Development Project, Peru [28]   157

The Conservation of Oku Mountain Forests for Wildlife,
Watershed, Medicinal Plants and Honey, Cameroon [29]   160

CHAPTER 4: TROPICAL FOREST RESTORATION   163

Agroforestry and Outreach Project, Haiti [30]   167

Growing Forest from Habitat Fragments in
Guanacaste National Park, Costa Rica [31]   170

Reforestation of Amazonian Bauxite Mines Using Native
Species, Brazil [32]   173

Rehabilitation of Damaged Ecosystems in the Amazon Basin,
Brazil [33]   176

Plan Bosque: Incentives for Planting and Tending
Trees, Ecuador [34]   177

Xiaoliang Water and Soil Conservation, China [35]   180

The Forest Villages, Thailand [36]   182

Soil Conservation on Steep Tropical Slopes, Philippines [37] 184

Village Forest Project, Uganda [38]   187

Notes   191

Suggested Reading   201

Index   203

Also Available from Island Press   209

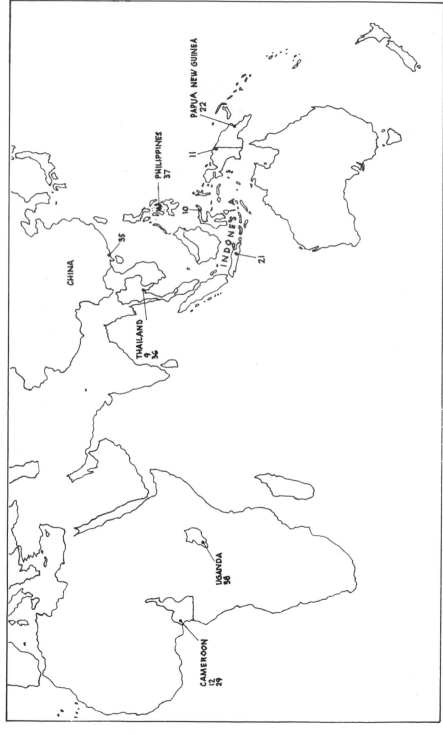

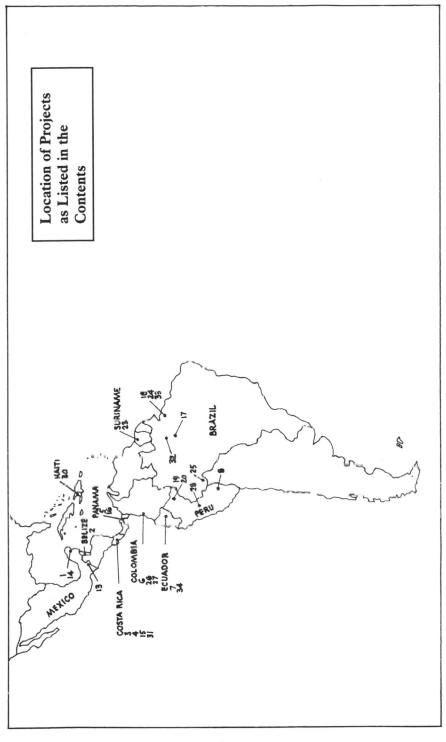

Location of Projects as Listed in the Contents

ix

BEYOND DESTRUCTION, SUCCESS

by Michael H. Robinson

At times it seems as though everyone is aware of the imminent threat to biodiversity and the destruction of the great tropical rainforests. These are issues that have captured the imagination of citizens in many countries, and this interest is being translated into concern. Conferences, workshops, seminars and protests on rainforest destruction and related environmental/biological problems have mushroomed. Even museums and zoos have developed exhibits highlighting these problems. In the circumstances, it is easy to be buoyed up by optimism and to think that for once in human history reason will prevail and we will stop the destruction before it is too late. Some people believe that we have only to educate Third World politicians, developers and peasants and all will be well. Indigenous conservation organizations are readily funded from outside to fulfill this fond hope. A moment's reflection shows that such optimism is unwarranted. If rainforests were being destroyed because of stupidity or ignorance, a flush of enlightenment could conceivably lead to reform. But this is not the case; the rainforests are being destroyed not out of ignorance or stupidity but largely because of poverty and greed. These forces have not, historically speaking, been susceptible to the ameliorating effects of reason and logic. In fact, it is one of the tragedies of the philosophy of Enlightenment that it postulated that ignorance was the main cause of the evils that beset the human condition. This pathetic fallacy has diverted efforts for too long. We can only act to prevent the worst effects of the crisis in the tropics if we identify its real causes and then look for clues to solutions.

I have recently re-read Mayhew's great classic on the condition of London's poor during Victorian times.[1] This has provided me with what I think is an important parallel. At the height of Britain's expansion as an industrial power, London was an awful example of disorganization, poverty, abject misery and crime. The condition of London's poor a little more than a hundred years ago was closely similar to that of substantial numbers of tropical peoples today. I have visited a variety of cities and towns in more than twenty tropical Third World nations. Many of them are Mayhew plus television. It was intolerable that such conditions should exist alongside extreme wealth in Victoria's day and it is equally intolerable that widespread poverty, privation, undernourishment, unhealthy and insanitary housing conditions and general backwardness should now afflict so many people in the tropics. They need accelerated economic development on a grand scale if the international polarization of wealth and poverty is to be abolished. There is no question but that the destructive pressures on tropical environments result from the attempts of peoples and governments to achieve economic equivalence with the developed world. We are thus faced with a familiar picture in which the inexorable short-term demands for food and commodities menace the long-term prospects for sustainable development and the conservation of global assets. The problem is essentially political/economic (not political *and* economic because the two are inseparable).

So why are we scientists involving ourselves in a rainforest/ biodiversity conservation campaign? We have no particular expertise in the area of politics—probably the opposite. We do, however, bring two things to bear on the problems. First, I think that we have a uniquely acute awareness of the probable consequences of our present actions. I am sure that our sense of urgency about the nature of the crisis is totally unrivalled. Secondly, we should have some insights into feasible alternatives to present systems of resource exploitation, since the problems are essentially concerned with increasing the yield of natural systems—of foods, fuel and raw materials. These two reasons seem to me ample justification for entering the fray, for throwing down the gauntlet to the developers.

It is no exaggeration to say that the biologists were the first to recognize the existence of the biodiversity crisis. The central expression of this impending tragedy, against which all its other manifestations pale into insignificance, is the destruction of that immeasurably diverse system, the rainforest. Evolutionary biologists in particular think in terms of the ebb and flow of phylogeny, extinction and speciation, and their patterns through time. We tend to think in vast paleontological time scales against which the life of our own species is ephemeral. We are aware that we are facing the first human-created extinction cycle that is on the same scale as the major geological events of the past, but condensed into an incomparably smaller span of time. This is not hyperbole, not even hypothesis, but imminent actuality. As evolutionary biologists we are appalled at the spoilage of species that have taken so long to reach their present state. This parallels the wanton destruction of the masterpieces of human culture that happens in war. It is offensive that the products of an infinite number of genetic changes, accumulated over an immense period of time, should be endangered by a short burst of economic development spurred by short-term "necessity." We are shocked that this immense species-loss should take place even before the organisms concerned have been appreciated for the complexity of their adaptations, the beauty of their form, the intricacy of their interactions and inter-relationships, their role in the global bio-sphere, or their potential usefulness to *Homo sapiens*. We alone probably fully appreciate the irreplaceability of the plants and animals that will soon be gone for ever. In stark summary: we are destroying irreplaceable species on an unprecedented scale without regard for their potential economic, aesthetic or biological signifi-cance. Should anyone think that the scale of impending destruction has been exaggerated, he or she should remember that the worst estimates were made *before* Erwin increased our enumeration of animal species on earth by factors of ten or twenty.[2] It should also be taken into account that it is in the tropics that Erwin's new estimates add the great majority of new species (from 1.5 million species of insects to 30 million species!).

The lesson about the danger to the tropics has been well publi-cized, thanks to the biologists. What about the alternatives to des-truction? There is no question that most of the attempts to grow plants and animals in the tropics have been geared to producing food for subsistence and commodities for cash. Trees have been harvested for foreign exchange as though they were minerals to be mined. They have been cut for building materials and burned in increasing quantities for fuel. In addition, forests are cleared to produce land for agriculture and cattle ranching. Agriculture on cleared land has ranged from swidden to agribusiness, and animal production has ranged from small-scale to the *latifundio*. Whatever its scale or its function and purpose, tropical natural resource production has been overwhelmingly based on methods (and science) developed in, and appropriate to, the north temperate region. It has involved, almost exclusively, non-indigenous species of plants and animals. In fact, most of applied tropical science has been devoted to making the Third World a better producer of commodities for the developed world's conspicuous consumption. That is its most frightening aspect. *In situ* research into the value of local methods, local plants and local animals is almost pitiful in scale. This is where we should be able to make suggestions, predictions and have insights.

It was with all this in mind that the conference that inspired this book was organized. It was organized by the National Zoological Park, a part of the Smithsonian Institution, with the authors of this volume, Judy Gradwohl and Russ Greenberg, as prime movers. The intention was to bring together as many people as possible with experience in, or insights into, methods by which forests could be preserved, in varying states from pristine to disturbed, in ways consistent with continuing economic advance for tropical peoples. As you will appreciate, as you read the reports from all over the world, success **is** possible. Rays of hope shine in chinks through the general darkness.

We desperately need more time for infinitely more research. If the world expenditure on tropical biology could be increased by a hundredfold it would still be a tiny fraction of the world expenditure on defense; it would be infinitely less than the value of even civil airplanes produced in the USA in one year. Increased research could change the whole future of our planet. It could dwarf space exploration in its contribution to "ground truth." Such research is possible, the problems are solvable; and there are even enough bright-eyed young biologists around to do it. Science **can** work for our benefit. Give us the jobs and we will produce the tools. But it will take time even if we start tomorrow. In the meantime the forests and the species are going. To buy time for the research we need a moratorium on destruction.

What are the possibilities? I think there are essentially three possible scenarios. Most positive is to establish a moratorium on forest destruction, or a lesser option, to conduct environmental triage. The worst case scenario is the *status quo*; if we continue the present destruction the question is not **if** the rainforest will disappear but when. The hopeful and relatively hopeful scenarios depend on solving a major set of problems. In order to halt destruction some means has to be found to continue to advance economic standards in the tropics at the same time that further deforestation is arrested.

This is an immense task. How can we relax the pressure on land clearance and sustain development? Even in Jefferson's day it was more economical in the short term to clear new land than to increase the productivity of old land. There may be some solutions. Rubinoff has suggested levying money from the developed countries to subsidize forest preservation in the tropics.[3] This would provide a cash substitute for the products of the land to be cleared. It might even provide the income necessary to buy the agricultural surpluses of the developed world. No-one at the political level has even picked up this ball, never mind run with it. What then? Perhaps if we can't save most of the forest we can do environmental triage—abandon the hopeless cases, and concentrate on the best chances. The hopeful options are diagrammed, as I see them, in Figure 1 below. But even triage is not meant as a substitute for the research and species-

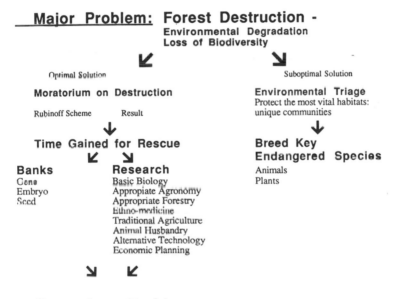

Major Problem: Forest Destruction -
Environmental Degradation
Loss of Biodiversity

Optimal Solution

Moratorium on Destruction

Rubinoff Scheme   Result

Time Gained for Rescue

Banks   **Research**
Gene   Basic Biology
Embryo   Appropiate Agronomy
Seed   Appropriate Forestry
                 Ethno-medicine
                 Traditional Agriculture
                 Animal Husbandry
                 Alternative Technology
                 Economic Planning

Suboptimal Solution

Environmental Triage
Protect the most vital habitats:
unique communities

**Breed Key
Endangered Species**
Animals
Plants

Secondary Problems

Provide Sufficient Fuel
biogas
plantations
utilization of new species
efficient stoves
alternative energy sources

Provide Sufficient Food
alternatives to destruction
better use of existing land
use of new species of plants/animals
agriculture
land reform

**Develop Economy
Non-Destructively**
New relations with developed world, reforestation
particularly aid programs
rational exploitation
economic reforms

banking proposed in the best-case scenario. A massive research program is the key to it all.

There **is** a model for the research effort involved. I have just read Rhodes's brilliantly fascinating book *The Making of the Atomic Bomb*.[4] This account of the achievements of a relative handful of scientists, given almost infinite resources and logistic support, is salutary. Despite misdirection, bureaucratic inertia, conflict and sheer organizational stupidity, they achieved their objectives in an "impossible" time-span. We need now the equivalent of a Manhattan Project for the future of life on earth; the threat may be ultimately as dire as that which faced the physicists of that time. We know there is hope because the conference and this book show us examples.

FOREWORD

Tropical deforestation is one of the major environmental problems now facing us. The tropics are being converted rapidly from expanses of pristine forest that support an enormous diversity of wildlife and plants and a richness of human cultures into vast wastelands that support a few tough, fire-resistant weeds and perhaps some cattle, while people scrounge for food and fuelwood from the newly-degraded soils and sparse shrubbery.

In the face of this grim picture, solutions seem rare. Deforestation, because of its global nature and its complex connections with Third World poverty, appears relentless and irreversible. Given such an apparent "manifest destiny" there seems to be little ground for optimism. Yet throughout the tropics a growing number of people are undertaking projects specifically designed to promote the wise use and preservation of remaining forest lands. It is our belief that by building on their efforts tropical deforestation can be stemmed and possibly even reversed.

This book grew out of a conference sponsored by the Smithsonian Institution, Friends of the National Zoo and the World Wildlife Fund-US in December 1985. The conference was organized to bring together people involved in developing positive approaches to tropical forest conservation. As time passed we learned about more and more people working on projects around the world to tackle the problem of deforestation. Their contributions—the story of their projects and of projects elsewhere—have provided the substance of this book. While we have modified contributions and written some case studies from background materials in the interests of presenting a common voice, the contributors and others who gave us their time did more than just assist: they are the true authors of this volume.

It has become obvious that tropical forest conservation presents an unprecedented environmental problem. The conservation effort cannot focus solely on establishing forest reserves (although large reserves are a necessary part of any conservation strategy). Because the clearing of forests is symptomatic of a pattern of uncontrolled and destructive land use, it is unlikely that parkland can be protected within a landscape that is otherwise completely overexploited. Unless the surrounding areas are developed in a sustainable manner and local communities are sympathetic to the protection of woodlots, greenbelts and larger parks, forest conservation will probably not succeed. Tropical forest conservation projects, therefore, must include not only forest reserves, but innovative approaches to economic development as well.

Several important caveats should be made at this point. Although we present a number of development projects that could provide models for the sustainable use of tropical forests, it is meaningless to discuss any technology or approach without also considering the ecological, social and economic context in which it is applied. One project presented here, for example, involves cattle ranching—normally highly damaging to the environment—on the Atlantic slope of Costa Rica. Unusually, its aim is to increase the efficiency of pasture use by converting from beef to dairy cattle and using the manure for other crops, thus reducing the need for members of a local community to clear more forest for pasture. But if investment and tax benefits continue to favor those who clear forest land for pasture, as in some countries, then this new approach may not reduce deforestation.

One other example should suffice. Clear-cutting tropical forest for paper pulp is generally destructive and uneconomic. The Pulpapel project at the Bajo Calima Concession in Colombia, however, has a number of aspects of design and implementation that make it an attractive model for tropical forest development. Its logging practices cause minimal soil compaction and nutrient loss from the ecosystem, and the overall pattern of cutting protects potential seed sources. Extensive studies have shown that profitable logging can be conducted in this manner and leave tracts of land that support rich forest regrowth. Such land may not be as good as tracts of virgin forest for protecting all wildlife, but it probably affords better habitats for animals and birds and superior watershed protection than many other possible land uses.

Our second caveat concerns the use of the term "successful".

Success in conservation is an elusive property. With a change in the political structure or economy of a region, any sustainable management scheme can be disrupted and any reserve can be colonized. At best, the schemes and plans outlined in this book may be considered promising. The road to tropical forest conservation is already strewn with the corpses of elegant management schemes and paper parks that protect denuded hillsides. In preparing this book, we have found that many of the most interesting and innovative projects are quite recent. Whether they involve park protection in presently undeveloped areas, or agroforestry with trees that require fifteen to thirty years to mature, their success can only be measured years down the road. We have included such projects if they showed early signs of commitment to long-term success. But we advise the reader to check on the status of any project in the months and years after this volume is published.

Deforestation critically affects forests of all types. Of necessity we have had to restrict our discussion to lowland moist tropical forests. It is difficult enough to generalize from one project to another in similar ecosystems without trying to make comparisons between projects in areas with dissimilar climates and soils. However, although we focus on tropical moist forests, we have purposefully kept our definition broad so that it could encompass some important projects which would have been excluded by a stricter classification.

The project case studies are intended to be short and informative but not exhaustive. They give the reader a general picture of the goals and accomplishments, as well as of the pitfalls and problems, of a project. In some instances, we have described promising research or ecologically-sound traditional and indigenous practices in addition to actual conservation projects. For both case studies and descriptions of research, we have avoided thorough analyses and critiques; these can be provided by the contacts and references listed at the end of the book. Clearly this volume is far from comprehensive; it is meant to spark debate and further research rather than to provide definitive answers. We believe, however, that the challenge of saving the tropical forests is so great that any attempt, however limited, to address the problem should be brought to public attention.

ACKNOWLEDGMENTS

Our deepest thanks go to Michael Robinson for his unwavering support of this book. We are also grateful to Resources for the Future for the grant that supported the research and writing of this volume. The 1985 conference, Tropical Forest Conservation, was sponsored by the Smithsonian Institution, Friends of the National Zoo and World Wildlife Fund-US.

For help with the essential tasks of editing, organizing and researching background material, we are particularly indebted to George Angehr, Eric Carlson and Elliott Gimble.

We relied heavily upon the generous assistance of a large number of experts in the field. In addition to the contributors to the case studies were those who participated in the conference, provided leads and information about the projects or reviewed materials. All gave their valuable effort and time:

Sheldon Annis
Peter Ashton
James Baird
James Barborak
Robert Blake
Barbara Bramble
E.F. Bruenig
James Burchfield
William Burley
Robert Bushbacker
Martha Cappelletti
Mac Chapin
Jason Clay
Walter Corson
Julie Denslow

David Deppner
Lou Ann Dietz
Gretchen Ellsworth
Terry Erwin
Julian Evans
John Frechoine
Curtis Freese
Carl Gallegos
Dennis Glick
William Gregg
Malcolm Hadley
Lawrence Hamilton
Marea Hatziolis
Randy Hayes
Susanna Hecht

Robert Hoage
Kathryn Hunter
William Hyde
Eric Hyman
Frazier Kellogg
Peggy King
John Michael Kramer
Molly Kux
David Lamb
Gerald Liberman
Thomas Lovejoy
Diane Lowrie
James Lynch
Theodore MacDonald
Craig MacFarland
Cynthia Mackie
Joan Martin-Brown
Thomas McShane
Cynthia McVay
Eugene Morton
John Muench
Norman Myers
Richard Norgaard
Roelof Oldeman
John Palmer
Francis Putz

Herbert Raffaele
Timothy Resch
Bruce Rich
Phyllis Rubin
Ira Rubinoff
Roberta Rubinoff
Katheryn Satterson
R. Sakumar
Jeffrey Sayer
William Siegel
Ross Simons
John Spears
Frances Spivy-Weber
Thomas Stoel
Roger Stone
Thomas Struhsacker
Cindy Taft
John Vandermeer
Martha Van der Voort
Craig Van Note
A.P. Vayda
Richard Warner
Charles Wendt
T.C. Whitmore
Michael Wright

Administrative and clerical support was provided by James Fitz-patrick, Gail Hill, Annette Miller, Marty Rogers, Barbara Russell and Monica Valley. Lois Sloan produced beautiful line drawings in very little time, and the graphics department at the National Zoological Park provided the maps for chapter headings. Finally, we are grateful to Diana Page, Sarah Stewart and Neil Middleton of IIED who skillfully steered the manuscript through the complex process of publication.

The opinions expressed in this book reflect the views of the authors and not necessarily those of the Smithsonian Institution, the International Institute for Environment and Development or Resources for the Future.

PART I
THE SPECTER OF DEFORESTATION

THE SPECTER OF DEFORESTATION

Since the advent of agriculture, the forests of the world have been cleared to make way for pasture and cultivated fields. When the Romans arrived in Britain nearly two thousand years ago, much of the land was covered with broad-leafed forests. Today, the island has been stripped of most of its woodlands, and the land is used to graze animals, grow crops and house people. The pastoral scenery of the countryside, a land for relaxation through walking or riding, is the outcome of wholescale conversion of forests. The same story of deforestation can be seen throughout the world, whether it be the hills denuded in biblical times in the Middle East or the farmland cleared only a few generations ago in the mid-west and Atlantic seaboard of the United States. In a few areas in the middle latitudes, most notably the eastern United States, a period of intense clearing of woods has been followed by the abandonment of agriculture and recovery of forests. But on the whole, within the last thousand years there has been a substantial new loss of forest land.

The effects of this historical deforestation in temperate and mediterranean areas have been mixed. Obviously the plants and animals associated with forests have declined simply as a result of the reduction in cover, and in some cases because the remaining forest did not provide the same habitat of an equivalent area of continuous woods. The long-term benefits to human populations have varied as well. Certainly some people have cleared their forests in favor of productive agricultural land. Other areas which have lost forest, notably the Sahel and southern China, have succumbed to erosion and desertification. Even where forest was logged, allowed to regenerate and eventually managed for sustained use of wood products, there have been profound ecological changes. Many wood-producing regions are managed in extensive monocultures with relatively short rotations. This is especially true in Europe and the United States.

Despite the historical precedents for extensive deforestation, at no time in history has the rate of deforestation approached what we are seeing as we enter the 1990s. Most unexploited forests are in the humid tropics, with a majority in the Amazon and Congo Basins, and it is in the humid tropics that the frontier of forest clearing can be found today.

Besides the intensity of clearing and burning in the tropics today, several other features distinguish the current wave of deforestation from what has gone before. The deforestation frontier is truly global, occurring in at least forty-five countries around the equator. More importantly, tropical forest lands, particularly those that have remained sparsely settled, cannot easily support the same types of intense and widescale use that characterized agricultural lands in the mid-latitudes. As the wave of colonization and development clears new areas, it creates in its wake a land often thoroughly wasted for any meaningful human use. In some areas forest clearing is followed by large tracts of monocultures for agricultural exports such as beef, oil palm, sugarcane, bananas or citrus in a bleak agro-industrialized landscape. But even in these productive regions little food is produced for landless peasants and the urban unemployed.

ECOLOGICAL SENSITIVITY OF TROPICAL FOREST LANDS

Lowland tropical forest land is difficult to exploit for several ecological reasons. One of the most important factors is the nature of tropical soils, at least those soils that underlie most of the remaining unexploited forestlands. In many areas of tropical forest, the soils themselves are relatively free of nutrients. This is particularly true of the iron-rich soils that can be found in the Amazon and Congo Basins. They lie atop ancient parent material that has undergone millions of years of weathering. High temperatures and rainfall throughout the year encourage leaching of chemical nutrients from the soil. With a few exceptions, the nutrients in a tropical forest ecosystem are mostly in the plants themselves. When plants die and are attacked by organisms, the nutrients that are slowly released are trapped by fungus, the main agents of decay, and delivered to the roots of the trees and shrubs through complex symbiotic interactions. If the forest is felled and burned for agriculture, the nutrients are rapidly flushed into the soil, where they are available for crops for one or, at best, a few rotations. Without the slow and complex mechanisms of nutrient recycling, these free ions are literally washed out of the soil. Vegetative cover is reduced if the soil is compacted through mechanical clearing or subsequent grazing, and the exposed soil is eroded by large raindrops pounding its surface—a problem which is particularly acute on hillsides.

Not all tropical soils fit this pattern. Some are quite rich, such as those found on recently formed volcanic rock and floodplain soils whose nutrients are replenished through annual flooding. In general, areas with these soils, like the floodplains of the great Amazon, Mekong and Congo Rivers, and the areas of young volcanic soils such as Java and parts of Central America, are the first to be settled. Most of the best agricultural regions of the lowland tropics have already been developed and much of what is left is land that is truly marginal.

BARRIERS TO REGROWTH

In areas with a wealth of second-growth forest such as the eastern United States, it is easy to be complacent about the ability of forest to return after it has been cut down. The future ability of moist tropical forest to regrow on land that has been logged, cleared for agriculture or grazed, is by no means so assured. If the clearing remains close to forest and little damage is done to the soils, a forest can come back with remarkable speed. Only a few centuries ago, which, from the perspective of a large forest tree, is not long, much of the Meso-American forests were cleared, often very extensively.[1] Forests there today appear untouched although we can see the effects of human involvement upon close examination of the flora. A more recent example of abundant regrowth can be seen around the Panama Canal. The area surrounding the Canal was largely cleared of forest during construction from 1904 to 1915, but it was allowed to regrow. Now this forest, although recognizably different in flora and stature from old growth forest, is one of the most accessible examples of moist tropical forest in Central America.

But one only need travel west from the Canal along the Pan American Highway to appreciate less optimistic prospects for forest regeneration. For several hundred miles, land that was once covered by tropical forest is now sparse savanna made up of grasses and fire-resistant shrubs. The land has been largely abandoned, yet forest will not grow. This is not to single out the Panamanian countryside: in New Guinea, Kunai grasslands cover the once-forested slopes, while in the Amazon Basin African grasses and shrubs dominate the abandoned cattle pastures. In some areas, such as Haiti, even stubborn grasses are sparse and bare mineral soil and bedrock dominates the landscape. Where deforestation occurred over the centuries, such as in the Atlantic coastal region of Brazil and the mountains of southern China, there is little or no original forest to supply the seeds for forest recovery.

Contrary to the popular image of tropics covered with relentlessly growing jungle, tropical forests seem distincly unable to reclaim land once it has been cleared and used for agriculture. There are a number of reasons for this:

- Clearing and burning is likely to release most of the nutrients into poor tropical soils, which are then leached out by warm tropical rains.

- The process of forest regeneration is complex and the seedlings of hardwoods found in forest often tolerate only a narrow range of humidity and light conditions and cannot, therefore, grow in open areas. Under natural conditions a succession of different species, starting with a colonizer, provide this environment.

- Tropical forest seeds are most often spread by animals, sometimes by relatively few species. This means that a seed source has to be near the clearing and that the disperser populations must be able to withstand drastic habitat changes.

- Forest trees are often pollinated by means of animals, and breeding systems such as dioecy (where male and female flowers are on different plants) are common. These plants require cross pollination between individuals, and this means that they must grow in close proximity, with the proper pollinators in the area.

- Both pollination and dispersal are made more difficult by the relative rarity of tropical forest species. This rarity is a natural consequence of diversity, since tropical forests, unlike many temperate forests, are not usually dominated by a small number of species.

- Trees may require specific symbiotic relationships with fungi known as micorrhizae.
- There is a collection of ungerminated seeds in tropical soil known as a seed bank, but it is unclear how long seeds remain viable. The degree to which seeds can remain viable in tropical soils in the face of insect and fungal attack varies greatly between species. Further, the conditions found in large human-generated clearings are often too stressful for seeds to survive.
- Tough, browse and fire-resistant grasses and shrubs take hold after long periods of burning and grazing. These plants can effectivly block forest regrowth.[2]

Given these constraints, it can be seen that it "is easy to keep a good forest down."[3] The belief in the ability of forests to regenerate quickly to their original state is unfounded. We will now examine what happens when forest land is developed in various ways.

ECOLOGICAL CONSTRAINTS ON SUSTAINED FORESTRY

Even in countries in the north temperate zone, most forest land is not protected as pristine wilderness, but is managed for "sustained" timber production and other purposes. Management of forests can vary considerably, but it often involves actions that change the composition of the forest. This is accomplished through planting or seeding desired trees and clearing and thinning-out less economically interesting trees. In addition, the age structure of managed forest is usually radically altered so that large stands tend to be even-aged. Extreme examples of this type of management can be found in the stands of young pine grown for paper pulp in the northern and southeastern United States. Sustained management of forests still provides some economic incentive to maintain some forest cover; the US, for example, has more forest cover now than it did fifty years ago. Its forests are threatened only locally, in areas of suburbanization (this encroachment of the suburbs has, however, resulted in a recent decline in total forest cover from 309 to 290 million hectares). Long-term US planning for forestry has not always been practiced. The development of the American frontier was a period of reckless logging that led to severely eroded hillsides and the loss of topsoil. There is very little old-growth "virgin" forest left anywhere and modern US forestry is being practiced on the regrowth following the earlier period of rampant exploitation.

The obstacles to establishing a sustainable management scheme for tropical forests, while not unsurmountable, are far more imposing than those for temperate forests. Tropical forests are far less forgiving of human intervention. We cannot assume after the first wave of logging and settlement either that more ecologically sound practices will be put in place or that, even if such practices are begun, tropical forests will recover.

Foremost in the problems of conducting sustainable forestry in moist tropical forests is the sensitivity of their soils to disturbance. With increasing mechanization of the harvesting and extraction of timber have come high rates of damage to the non-target forest vegetation and soil.[4] Heavy machinery compacts the soil and causes ruts that encourage erosion. This is particularly true in areas of high rainfall and steep slopes, and many of the areas that are being logged today are in hill country, particularly in Southeast Asia. Clear-cutting for wood chips and pulp often leaves large clearings which expose soil to tremendous erosion by tropical rains.

The diversity that makes a rainforest a naturalist's dream can also, from the standpoint of long-term management, turn it into an economic nightmare. In Old World tropical forests an astounding number of tree species—over a hundred—can be found in an area as small as one and a half hectares. Most timber markets accept only a few of these species, which leaves tropical forest full of economically useless trees. This is not nearly so true of forests in Southeast Asia where many species, often from the same plant family, Dipterocarpaceae, are often closely related and can be marketed together. But even here, only about a hundred tree species out of a flora of a couple of thousand are harvested.[5]

In some areas, forests are harvested simply for a few, highly profitable trees such as mahogany and purple heartwood. On the face of it, such selective logging would seem to be an ideal compromise in the use of forest. It can involve minimal damage and could be economically profitable without the need to raze the woods entirely. However, the exploitation of such wood means that foresters must go farther and farther to find it. Multiple harvests then occur as new trees come into economic fashion.

Selective logging is usually the first economic use made of a forest when it becomes accessible by road. But it is seldom pursued for long and the cutting and harvesting can significantly harm a forest.

In addition to the impact on the soils, the creation of large roads and gaps allows fast-growing vines to cover the edges of the forest canopy, which discourages the regrowth of trees of marketable stature.

Careful harvesting practices can mitigate the damage to the forest. Yet as soon as the valuable trees are extracted, the value of the forest immediately declines. Without any management, most forests do not regenerate the valuable species rapidly enough to allow selective logging to compete successfully against other development possibilities.[6] The very roads that open up an area for selective logging commonly become the avenues for transient settlers looking for land, as well as for other types of development. If the timbering system is only marginally profitable or becomes unprofitable, it simply sets the stage for a more intense form of forest destruction.

Management schemes have been best developed in dipterocarp forests of Southeast Asia where a high proportion of trees are harvested. Much of the reproduction occurs in "mast years" when trees of many species simultaneously bear fruit. This tends to create even-aged cohorts, and makes harvesting easier and more economic. However, even the classic "working" silvicultural system, the Malaysian Uniform System and its derivatives, are no longer commonly practiced because cutting for timber has given way to other uses of the forest, notably rubber and oil palm production. Current management systems adapted to hill dipterocarp forest have produced mixed results.[7]

In the long run, the removal of vegetation and hence nutrients may limit the degree to which any forest management scheme is sustainable. Since a greater proportion of nutrients are generally stored in the vegetation of a tropical forest than in other types of forest, it is reasonable to expect that the effect of nutrient transport will be that much greater.[8] This can be ameliorated in several ways: young trees of desirable species can be left in place and their competition removed by poisoning; logs can be cut up on the site, slash left to decompose and fire excluded from the site so that nutrients are then released slowly and continuously. But despite the use of such methods it is uncertain whether a managed forest can survive because of loss of nutrients. Its decline may take only a few cutting cycles, but if the cycles last thirty or more years, the actual effect will not be noticed for sixty to a hundred years.

ECOLOGICAL CONSTRAINTS ON SUSTAINABLE AGRICULTURE

In areas where forests have been cleared, cannot crops be grown to feed the population or to provide earnings from export agriculture? This might be thought a reasonable alternative, but agricultural development in former forest areas is constrained by two key ecological factors: soil and nutrient loss, and pests and pathogens which attack low diversity agro-ecosystems. Both of these act most strongly against large-scale single crops. The immediate causes of nutrient and soil loss, as we have seen, arise because at least two-thirds of the soils in tropical moist forests are thin and poor in nutrients.[9] Plant cover prevents the soil from washing away in torrential downpours, while the speed of vegetable decomposition and the fungi which aid this process insure a continuous supply of water soluble nutrients protected from leaching. Long-term tilling of tropical soil leads to the loss of topsoil and of soil nutrients. In order to maintain productivity, mulching, the use of careful cultivating techniques and long fallow periods are required. Otherwise, as in a number of agricultural systems, the soil must be supplemented with large quantities of artificial fertilizers.[10]

Much tropical land is devoted to crops or plantations of single species for export to the industrialized countries. This type of development avoids some of the pitfalls associated with poor soils because the crops are planted on the limited areas of relatively rich soils. Rice-paddy agriculture, for example, can be found on the alluvial soils of the great rivers of Southeast Asia, while citrus groves are often located on the streamsides in Belize and the savannas of Brazil. Coffee is commonly grown on volcanic soils, and palm oil, cacao, coffee and tea plantations are being established at a rapid rate in the Amazonian slope of the Andean foothills. Corporations and individuals with these large land holdings often only plant a portion of their land and hold the areas of poor soil in fallow.[11]

Aside from the major questions of social policy raised by this type of land use, however, major ecological considerations remain. Monocultures are more vulnerable to pests and weeds than traditional mixed-crop systems.[12] An entire agricultural economy can be threatened by a single insect pest or pathogen. In Indonesia, for example, infestations by brown leaf hoppers of monocultures of single genetic strains of rice have caused millions of dollars of damage. Single-crop agriculture is also usually aimed at export markets. This puts developing tropical forest countries in potential competition with one another as they all seek to sell the same commodities. Prices in commodity markets fluctuate considerably and farmers respond to these fluctuations by varying the amount of land planted with that particular crop. Crop management is therefore not dictated by local ecological considerations, nor by local patterns of consumption, but by the vagaries of the world market. Perhaps more critically, one-crop agriculture often depends upon large amounts of fertilizers and pesticides which must be bought with foreign exchange. Farmers' profits rise and fall with these prices and in a poor economy the productivity of intensive commercial agriculture can decline dramatically.

HOW FAST ARE THE FORESTS DISAPPEARING?

Tropical forests have always been limited in number and size, at least in recent geologic time. This is because moist or wet forests are found on portions of the landmass between the tropics of Cancer and Capricorn, a region of the earth's surface where there is relatively little continental land. The major regions of tropical rain forest are in South America, Africa and the large islands of Southeast Asia. Three tropical forest countries—Brazil, Zaire and Indonesia—contain nearly fifty percent of the world's closed-canopy tropical broadleaved forests. These countries have only slightly more than half the forest cover of the Big Three temperate forest countries, the USSR, USA and Canada (580 versus 1,650 million hectares, according to World Resources 1986).[13] Most of the remaining tropical forest (57 percent) is found in the New World, another 19 percent in Africa and 24 percent in Australasia.

It has been estimated that there may have been 1.6 billion hectares of tropical moist and wet forest before widescale human-caused deforestation began in earnest. This area has already been reduced to about 1.1 billion hectares.[14] These figures are only estimates and are based upon data from the 1970s; we really do not have an accurate idea of how much forest remains. What we do know is that deforestation is now proceeding at an annual rate of at least one percent—up to two percent if areas under selective logging are included—and that, extrapolating from these figures, twenty to forty hectares are disappearing every minute. In some countries the situation is even

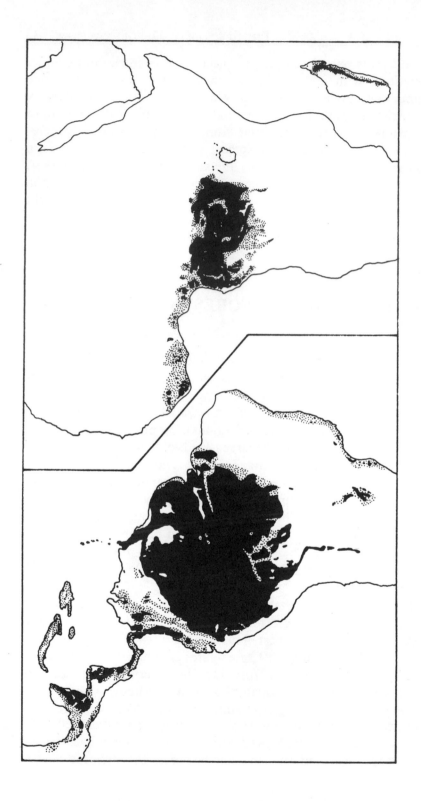

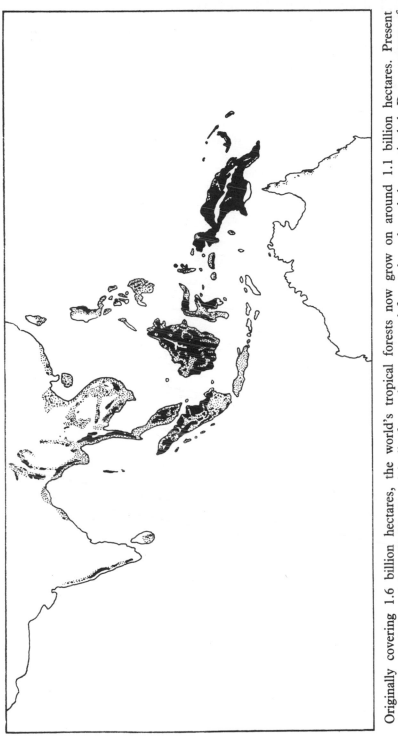

Originally covering 1.6 billion hectares, the world's tropical forests now grow on around 1.1 billion hectares. Present distribution is indicated in solid black; originally forested areas now deforested or degraded are stippled. Data courtesy of *Tropical Rainforests: A Disappearing Treasure*, Smithsonian Institution.

worse. In the West African countries of the Ivory Coast and Nigeria, for example, the rate of deforestation is as high as five to six percent a year. Other hot spots, as Table 1 shows, include El Salvador (3.3 percent), Thailand (2.9 percent) and Costa Rica (4.0 percent).

Table 1. Tropical Deforestation Rates in Selected Countries

	% Annual Deforestation	Forest Cover (1000 Km²)	% Logged or Managed
Central America			
Costa Rica	4.0	16	46
El Salvador	3.3	1	18
Belize	0.7	14	66
Panama	0.7	42	19
South America			
Ecuador	2.4	142	8
Colombia	1.8	464	1
Brazil	0.4	3575	3
Suriname	0.4	148	2
Africa			
Zaire	0.2	1056	4
Cameroon	0.5	180	55
Ghana	1.3	17	78
Nigeria	5.0	60	43
Asia			
Indonesia	0.5	1139	31
Papua New Guinea	0.6	342	1
Malaysia	1.2	201	22
Thailand	2.9	83	0

NOTE: The FAO figures used here tend to be conservative because only agricultural development is considered deforestation. We have therefore included FAO estimates of the amount of logged-over or "managed" forest land. Logged-over land can vary in the intensity and age of the timber operation but includes much highly degraded forests. "Managed" lands includes those lands slated for logging operations and apparently implies nothing about the sustainability of the operations. Other estimates can be found in N. Myers, *Conversion of Tropical Moist Forests* (Washington, DC: National Academy of Sciences, 1980).

SOURCE: J.P. Lanly, *Tropical Forest Resources* (Rome: FAO, 1982).

These figures do not convey the fate of particular types of forest: the *caatinga* and *varzea* forests of the Amazon, the dry forests of the Pacific slope of Central America, the unique island flora and fauna of Madagascar and the West Indies, and the mid-elevation forests of the Andes are all rare and unique habitats experiencing high rates of conversion. Nor do the figures show local variations. Within some of the larger countries, such as Indonesia and Brazil, we can find areas that are showing explosive rates of deforestation because of particular development activities.[15] As in certain parts of the Amazon, where deforestation rates appear to be following an exponential, rather than linear, trajectory, the economic situation of a country or region can cause the pace of deforestation to quicken or slow down. Currently the worldwide rate of deforestation is on the increase.

TO WHAT HAS TROPICAL FOREST BEEN CONVERTED?

CUTTING FOR FUELWOOD AND TIMBER

In many tropical areas, particularly Africa and Southeast Asia, the first step in the conversion of tropical forests is opening up areas for logging. In addition much of the developing world relies on wood for fuel. In fact, over half of the wood harvested in developing countries is fuelwood and other domestic products (poles, fencing materials, etc.) for home consumption. However, fuelwood collection is not so much a problem in tropical moist forests as it is in relatively arid areas. Moist forests suffer more from selective logging for tropical hardwoods or, in some areas, the production of wood to fuel large-scale industries.

Tropical forest wood is converted to charcoal to fuel steel and brick factories, iron smelters and cement plants. The process is highly inefficient: one ton of charcoal requires four tons of tropical forest wood and will fuel an iron smelter for only five minutes. The Grande Carajas regional development program in the Eastern Amazon includes an ambitious charcoal-production scheme. It is projected to produce and use 1.1 million metric tons of charcoal annually, to run seven pig iron, two iron alloy and two cement factories. Although some of the demand for charcoal may be met by eucalyptus plantations, it is unlikely that natural forests will be spared.[16] Mining further exacerbates the problems. Forests are cleared to allow for mineral extraction, and conflicts with indigenous people over land use rights have resulted in deaths.

Logging operations cover five million hectares of moist tropical forest each year. Particularly hard hit are Malaysia, Indonesia, the Philippines, the Ivory Coast and Gabon, which together account for about eighty percent of the total volume of tropical hardwood exports.[17] Thirty countries are exporting tropical hardwoods, but at the current rate of exploitation this number may dwindle to ten within a generation.[18] A few countries such as Nigeria and Thailand have already passed the brink, and are importing more wood products than they export. Although tropical hardwoods earn their exporting countries about eight billion dollars in foreign exchange, a number of factors decrease the value of this commodity for anyone but the companies involved in importing and marketing the wood. Timber concession land is often state-owned and logs are exported before processing, so foreign corporations, which own the means of shipping and processing timber, often make the bulk of the value-added profit. This makes harvesting a relatively unprofitable business and encourages cutting corners on sound management to make money on the local level. To make matters worse, the price of raw timber has been dropping steadily over the past thirty years.[19] On the other hand, several countries such as the Ivory Coast, Ghana and Indonesia have taken steps to increase domestic involvement in processing and exportation of wood. Unfortunately, this may not improve matters, as much of the local wood milling is thought to be inefficient and wasteful of timber.[20]

Where does all this wood go? Tropical hardwoods are valued for both aesthetic and practical reasons. Many are resistant to decay and attack from termites and other wood-boring insects. The largest importer of tropical hardwoods is Japan, despite the fact that two-thirds of Japan is covered with forests. But often its wood is processed in third countries like Taiwan and South Korea where cheap labor and low taxes and tariffs make the import of logs and sawn boards less expensive.[21] The USA is the second largest importer following Japan, though Europe as a whole imports more. Ironically, the USA receives most of its tropical hardwoods as finished veneer and other products from Japan. Other uses for lighter tropical forest woods include pulp, wood chips, matches and even chopstick production.

Despite the damage to the forests, selective logging for hardwoods is often not counted as tropical deforestation. The most widely cited figures for deforestation, produced by the FAO in 1982, specifically exclude logged areas. In most cases, however, no sustainable forest management is employed; it is unlikely that forests will remain intact long after the initial cuts are made. Access roads for loggers open up the land to more permanent forest conversion.

COLONIZING THE FOREST FRONTIER

One of the greatest causes of deforestation is the wave of slash-and-burn agriculture that is pushing back the forest frontier. Denevan has estimated that as many as 300 million people might clear forest plots, including both fallow and primary forest land, each year, with each family of eight persons requiring one to five hectares per year.[22] Estimates on how much new forest land is involved are widely divergent, ranging from 7 million hectares to at least 20 million hectares each year.[23]

Much forest that is called virgin or primary has been subject to slash-and-burn agriculture for centuries, so it is not inevitable that slash-and-burn agriculture permanently turns forest into pasture. Indeed, on long rotations it is a sustainable land use system. However, the current wave of colonization is so tremendous that land is not allowed to lie fallow for the twenty to a hundred years required for the regeneration of soil and nutrients.

Although the pattern varies from region to region, the basic thrust —and ecological impact—of slash and burn is the same worldwide. A new forest area is opened up; new roads are built specifically to encourage colonization. Poor rural families, following the promise of abundant farmland, become homesteaders, usually with rights to farm public land as long as it is kept planted. The forest lot is cleared, first by cutting the vines, understory shrubs and saplings and then by cutting the canopy trees with machetes or chain saws. Vegetation is left to dry and the whole plot is burned. During this time, the forest frontier is like a smoky scene from an old war film. The pall of smoke hangs over immense areas, and ashes carried by the wind cover anything left exposed. Planting occurs just before or after the burning so that the root and grain crops can take advantage of the pulse of nutrients released from the fire. Typically after the first year the plot begins to decline seriously in productivity; weeds and pests appear. Most subsistence farmers also try to supplement their income with products collected from the forest, such as wood, fibers and foods, or work as wage laborers. But in areas of recent colonization, such as the Amazon, opportunities for wage labor have dwindled and nearby forests have been overexploited. These factors, coupled with the drop in crop productivity, result in widescale abandonment of farms. In some regions, particularly in the neo-tropics, land is then reclaimed by large ranch owners for cattle operations.

New forest lands act as a safety valve for the serious economic and social problems facing many developing countries. Colonization has the advantage from a government's point of view of providing new lands without true land reform.[24] It also settles people into politically sensitive areas, often near international boundaries, where pre-viously only a low density of indigenous people lived. And no matter how desperate life becomes in the barrios and countryside, a living—occasionally a good living—can be hacked out of the jungle.

The laws in many Third World countries promote the clearing of new forest lands, even lands that have been declared national parks or forest reserves. Under squatter laws land-use rights revert to the homesteader after one crop has been raised, regardless of its previous status.[25] In these circumstances little is needed to encourage the colonization of primary forest areas. If a road is built, migration will usually follow.

Several countries, most notably Indonesia, have instituted massive resettlements to less accessible forest lands. In Indonesia, 3.5 million peasants have been moved from the inner islands of Java, Bali and Madera to the outer islands of New Guinea (Irian Jaya), Timor, Sulawesi and Sumatra. Today over 48 million hectares of previously forested lands have been degraded by settlers in Indonesia. Re-settlement programs are expensive; much of the funding has come from international financial institutions and development agencies.[26] The transmigration plan originally envisioned moving eighty percent of the participants to primary forests over the next five years, with the hope that a total of 65 million people would be relocated over the next two decades. However, the loss of government revenue due to the worldwide decline of oil prices and the expense of the program have forced a reduction in planned government-sponsored resettle-ment.

Although not as organized as Indonesia in its resettlement pro-gram, the Brazilian government has made a concerted effort to colonize the outer reaches of the Amazon Basin through the con-struction of a road network.[27] Much of the massive colonization expected along the Amazonian highway system failed to materialize, but recent road projects have attracted large numbers of people into the western part of the Basin. Most recently, a road completed in Rondonia attracted over half a million people, and caused tremendous rates of forest clearing. In addition, colonization has created a new class of rural poor and the severe disruption of tribal peoples' life.[28]

Hydroelectric dams are another type of major development project that destroys tropical forest. Sometimes covering thousands of square kilometers, the process of filling the reservoirs can drown large tracts of forest, displace people and kill wildlife. Although dams are intended to provide inexpensive electricity, many are economic failures because of lack of environmental planning. Erosion of watersheds quickly fills reservoirs with silt and reduces the ultimate output and useful life of dams.

LARGE-SCALE AGRICULTURE

Throughout the tropics, much of the land on productive soils is held in large farms where it is used to produce commercial crops, primarily for export. Generally only one or two crops are dominant in a given region. For example, rubber, oil palm and cacao plantations are common in Southeast Asia, as are rice paddies. Central American countries have been called banana republics for good reason, as much of the lowland Caribbean has been under the economic control of US corporations for eighty or ninety years. Much of the land is not usually controlled by foreign or multinational corporations, but is owned by wealthy nationals. However, shipping, processing and the manufacture of agricultural inputs such as pesticides and fertilizers are generally owned by foreign companies. This system is of mutual benefit to wealthy landowners and multinationals alike.

This type of agriculture has two major effects on land use. First, ownership is often concentrated in the hands of a small minority. Statistics can be presented for any number of tropical countries, but they are generally similar, and, as Table 2 shows for Central America, highly skewed in favor of a few.

The story is the same in other countries: in non-Amazonian Brazil, 4.5 percent of the landowners have holdings larger than fifty hectares and so control 81 percent of the land; in Zaire the top one percent of the farms include 40 percent of the land; in Peru, two

Table 2. Distribution of Land in Central America

	0.4 ha.	0.4–2 ha.	2–20 ha.	20–324 ha.	324+ ha.
% farms	23	55	14	6	< 1
% land	< 1	10	16	35	38

SOURCE: T. Barry and D. Preusen, *The Central American Fact Book* (New York: Grove Press, 1986).

percent control 78 percent of the land.[29] These figures contrast sharply with those for developed countries. But even these statistics understate the problem because in many of these societies the majority of families are legally landless. In rural Brazil, for example, seventy percent of the families do not have secure land titles. For rural agrarian societies, land is wealth; landlessness can mean starvation.

Large-scale agricultural holdings usually include the best farmland. For example, in Honduras, a mere five percent of all landholdings contain two-thirds of the country's fertile land, and of this, two-thirds is used for cattle ranches.[30] Management of these prime lands is based on the export market, and with tax incentives it is often profitable for much of the land to be left fallow. In addition, large farms can be more wasteful of good agricultural land than small holdings.

Agribusiness is designed to maximize the profits of the landowners, not the economic or social benefits of the other local residents. It disrupts the local subsistence culture and replaces it with a wage labor economy. If agribusiness firms employed large numbers of landless peasants, it might help curb deforestation. But usually the jobs are few and highly seasonal. In the 1970s, unemployment in the sugar-growing regions of the Dominican Republic ranged between forty and sixty percent—a generous estimate, because the average duration of employment was only 135 days each year, during the harvest. In El Salvador prior to the war, unemployment reached forty to sixty percent while other Central American countries suffered unemployment rates of fifteen to almost fifty percent.[31] To survive, women garden or gather forest products when their men go off to work.[32] But this sustenance disappears when forests or second growth are removed from the landscape. Cattle ranching, so preva-

lent in Central and South America, epitomizes the system of agribusiness that deprives people of land and contributes little to the
production of local food or wealth.[33]

COMPLEX INTERACTIONS: A COLOMBIAN CASE STUDY

Most tropical forest countries have two parallel rural economies that
affect each other and the rate of deforestation. Recent developments
in rural Colombia illustrate this interrelationship.[34] With its 113
million hectares of cropland Colombia has tremendous agricultural
potential, yet much of the country lives in rural or urban poverty.
Urban poverty has grown quickly in recent decades partly as a result
of policies to develop industries based on inexpensive labor, and has
been accompanied by a large variety of social ills: drug production,
the clearing of new forest land, and emigration, including that of up
to a half a million people to the USA since 1959. Roughly a fourth of
the country is under some sort of cultivation, one-half of it in cattle
pasture. Cattle ranching in Colombia requires little by way of capital
expenditure compared with other forms of commercial agriculture,
but it is land-expensive, competes poorly on the international
markets, brings in little foreign exchange per unit of land, and
creates relatively few jobs (approximately 14 jobs per 1,000 hectares). For this, most farmers are pushed onto small holdings or have
no land at all. Three-quarters of the farms are ten hectares or less in
size and cover seven percent of the land, whereas 0.7 percent of the
farms are larger than eight hundred hectares and cover 41 percent of
the land.[35] Small farmers tend to move to steep hillsides where
cultivation causes disastrous levels of soil erosion.[36] Poor families are
encouraged to go the new frontiers and cut farmland from forest.

Newly opened lands are not always dominated by colonists with
small holdings. When 12,000 hectares of land opened up in the
eastern Andes from 1950 to 1973, 54 percent was devoted to ranches
and only 16 percent to crops. The rest of the land was left fallow.
Most of the land initially cleared by colonists was abandoned and
then incorporated into large ranches, or "latifundios."

Ironically, urbanization probably increased the pressure on forest
lands. Rice is the major commercial food crop grown on large
holdings in Colombia, primarily for the domestic market. Its production has been encouraged to provide an inexpensive source of food
for urban workers. Rice production is not only carried out on some
of the best farmland, but receives most of the agricultural credits.

Rice farming requires considerable inputs of fertilizer and pesticides which need to be purchased with precious foreign exchange. Small farmers cannot compete with the well-supported large growers and this forces families to move to new lands, grow more lucrative crops such as coca, or move into the cities. While this is a story whose particulars apply only to Columbia, the point is that deforestation is often an outgrowth of the entire pattern of post-colonial economic development of tropical countries.

OTHER PROBLEMS

THE GROWING HUMAN POPULATION

In 1987 the world's human population reached five billion, and we are adding more than eighty million people each year, mostly in the Third World. At this unprecedented rate, the demand for potable water, food and housing will increasingly tax the world's natural resources. Many groups, acknowledging population growth as a critical barrier to forest conservation, have incorporated birth control into their conservation programs. However, population growth is both a consequence and cause of rural poverty in tropical countries. In economies where the wealth of families depends upon how much land can be cleared and farmed, and how much labor can be performed, children are a fundamental asset. As long as large families are economically optimal and a cultural norm, population control cannot be effective. Nonetheless, a growing population does ultimately threaten rural resources such as forests. Even countries that emphasize social services, production of food and equal distribution of land and resources recognize that population growth causes an increase in rural poverty and makes social programs difficult to implement. The problem of population growth is exacerbated in tropical forest areas because natural growth is often augmented by immigration.

THE DEBT CRISIS

Large foreign debts incurred by tropical countries also play an important, if indirect, role in tropical deforestation.[37] Many countries borrowed heavily in the late 1970s in an attempt to offset the rising price of oil and to keep their economics growing. A combination of rising interest rates and global recession has made it next to impossible for most countries to pay off the loans. A one percent rise in US interest rates could add about eight billion dollars to the global debt bill by 1990.[38] Ever-rising export earnings are then needed to meet the staggering debt service payments. Brazil, for example, holds loans totalling over $100 billion and has had to spend nearly forty percent of its export earnings on debt payments. Many governments have few options beyond the rapid extraction of natural resources to raise capital, and the most readily available of these are tropical forest products such as timber, agricultural products, minerals and oil. Ironically, many of the loans contributing to the debt burden were for development projects such as roads and hydroelectric dams that resulted directly in forest clearance. Even deferments of loan payments can contribute to deforestation. Conditions for rescheduling payments usually include reductions in government spending, and environmental programs are usually among the first victims of cutbacks.

WAR AND WILDFIRES

War destroys tropical forest habitats and modern warfare has proved particularly lethal. During the Vietnam War, 44 percent of the country's rainforests were defoliated with herbicides, while 25 million bomb craters displaced three billion cubic meters of soil.[39] Even today, military activities threaten several areas of tropical moist forest, including the Caribbean slope of Central American and northern Burma. What makes warfare in tropical forests particularly ominous is that part of the strategy of counterinsurgency operations is the destruction of forests that could harbor guerillas.

Wildfires, although rare in rain forests, can also be destructive. These fires seem to result from logging operations in Southeast Asia and clearing for pasture and logging in the Amazon. The great fire of 1982–83 in Kalimantan, Indonesia, and Sabah, Malaysia, destroyed three million hectares. It occurred during an El Niño-induced drought, but slash from logging was one of the contributing factors.[40]

THE CONSEQUENCES OF DEFORESTATION

THE POSSIBILITIES

If there are few reliable data on the extent of tropical forests, there is even less information on the potential consequences of the widescale loss of this habitat. There has been much discussion about the effects of tropical deforestation on wildlife, quality of life for tropical forest people, local and global economies and the climate. But the nature of the problem is that most of these effects will not be felt until they are irreversible. By allowing deforestation to continue unchecked we are conducting an experiment of global proportions unparallelled in the history of natural events and science. Scientists have begun to study the myriad complex interactions and interdependencies that occur within tropical forests and can only speculate about the relationship between tropical forests and the rest of the global ecosystem. Even the most skeptical, however, will admit that some ties exist, and that we are tampering with a complex system. Among the varied predictions are:

- Loss of people and cultures whose way of life depends upon the forest, along with a loss of their knowledge;
- An increase in barren land and desertification in drier tropical regions;
- Regional decreases in rainfall, exacerbating desertification;
- Global increases in temperature due to a rise in atmospheric carbon, leading to a rise in sea levels;
- Extinctions of large numbers of plant and animal species, including the loss of important wildlife species and potentially important food and medicinal plants;
- Declines in temperate zone birds that migrate to the tropics;
- Increased exposure and erosion of soil;
- Loss of hydroelectric power potential; and
- An increasing downward cycle of rural poverty.

Although these predictions have been made by people working and thinking about the problems of deforestation, many of them are contentious. Some may overestimate the impact of deforestation, and some may give too little credence to it. Certainly there may be consequences that, based on our current knowledge, cannot be foretold.

Specific predictions about the effects of deforestation are difficult to make because we do not fully understand what comprises a tropical forest and the rules that govern how both tropical forest and tropical agro-ecosystems work. Even basic information on the pattern of land use in most tropical areas consists of estimates that were compiled almost a decade ago.[41] The data on deforestation rates are based largely on information provided by governments of tropical forest countries and only rarely has direct satellite imagery been analyzed to calculate rates of forest conversion.[42] Alarmingly few stations exist where long-term studies can be conducted on the ecology of tropical forests, and funding for basic studies is scarce. There have also been few hydrological studies of the effect of deforestation on specific watersheds, even when the future of a globally important structure like the Panama Canal is at stake. Basic information such as long-term weather data is difficult to obtain for many tropical forest sites. Keeping in mind that even the most dire predictions may be true, we will pursue a brief discussion of the most generally accepted effects of deforestation.

THREATS TO INDIGENOUS PEOPLE

All tropical forests have, at least until recently, supported tribal groups for thousands of years. These groups have a range of ways in which they exploit the forest, and most possess sophisticated ecological knowledge about the forest. A small number of groups such as the Mbuti (pygmies) in the Ituri forest in Zaire are hunter-gatherers, although the Efe pygmies in the northern part of the Ituri rely upon cultivated crops for over half of their caloric intake. They trade their labor in agricultural fields and forest products for this food. Many groups, such as the Bora of Peru, Kayapó of Brazil and Lacandon Maya of southern Mexico, practice shifting agriculture in addition to hunting and gathering forest products. The Dayak of Borneo maintain more permanent agricultural plots and fisheries along the rivers but also exploit resources in the upland forests. All of these groups depend upon the forest to a large degree for both material wealth and spiritual values. The destruction of tropical forests works in a number of ways to disturb, acculturate or decimate the original inhabitants. Conflicts with developers and colonists of forest land, and contact with diseases to which they have no resistance are by-products of deforestation which can be as disastrous as the loss of land for farming, gathering and hunting. Often tribal people live in small, widely scattered settlements or family groups, and have little political unity for facing or resisting a national government.

The development of the forest threatens the lands of the Dyak, the indigenous Malay or Proto-Malay people of Borneo.[43] In both Kalimantan and Sarawak, land claims of tribal groups in the forest are unclear and usually unprotected in the face of logging operations. Concessions are granted to corporations much more readily than they are to communal groups. In addition to usurping and degrading the forest upon which the Dyak depend, logging operations cause massive flooding and siltation of the rivers which destroys much of the local farming and fishing. The massive relocation programs, particularly in Indonesia, intentionally move people from the central islands into tribal areas in what some people think is a deliberate attempt to "Javanize" these remote areas.[44] In many areas indigenous people no longer practice traditional methods of resource exploitation. The basic social systems of tribal peoples are disrupted as they increasingly join the wage labor force. In most cases they are the lowest on the economic ladder and suffer from high rates of unemployment.

In most of Central America, Indians have ceased to live in traditional ways on forest land. In the Amazon the tribal population has been substantially reduced as the groups are driven into smaller and smaller refuges. There has been some effort by Brazil's Fundacao Nacional do Indio (FUNAI) to protect Indian lands from large-scale encroachment, but in areas experiencing massive colonization it is inevitable that the *colonos* searching for new land to cultivate will come into conflict with indigenous groups.[45]

As fewer tribal people engage in traditional practices, valuable knowledge gained over hundreds of years in tropical forests is being lost. From archeological studies it is possible to infer only the broadest outline of how forest resources were used and how farming was done. Recent studies of the existing practices of traditional systems have revealed remarkable levels of sophistication, much of which could be of benefit to modern agroforestry.

SOIL LOSS

Forest vegetation protects soil from the full force of tropical rain storms. The canopy receives the brunt of raindrops that are further broken up as they hit the understory foliage. Often, understory plants have long narrow "drip tips" at the ends of each leaf so that the impact of the rain is further reduced as it drips down to the forest floor. When forest is cleared, and particularly if the vegetative cover on the soil surface is removed, the bared earth receives the full brunt of intense cloudbursts and the thin topsoil is washed into nearby streams. This loss of soil causes two distinct problems. First, the loss of soil from hills makes it difficult to re-establish any sort of vegetative cover, either in the form of crops or trees to protect the watershed in the future. The second is that displaced soil washes into rivers and disrupts stream ecology and increases sediment loads in weir dams downstream. Erosion can be particularly devastating in areas of steep topography; it was recently found that 40 tons per hectare of soil was lost from the slopes of the Cauca Valley watershed in Colombia during a period of ten months.[46] Another study found that in an area of the Ivory Coast with slopes of only seven percent, forested areas lost about .03 tons of soil per hectare each year, but that cultivated slopes lost 90 tons per hectare, and bare ground lost 138 tons.[47] Resulting siltation threatens major dam projects such as Ambuklao in the Philippines and the Panama Canal.[48]

FLOODS AND DROUGHT

While reducing the water table, forests hold moisture in the ground and regulate its flow into streams.[49] Removal of natural forests, though usually increasing the total flow, results in water flowing into rivers in a pattern of alternating episodes of flood and drought. This causes severe problems for agriculture and other types of activities such as fishing, and can severely reduce the efficiency of irrigation and hydroelectric power. In tropical areas, rain is recycled into the atmosphere largely through transpiration of plants which can contribute to local atmospheric humidity and ultimately to rainfall. In the Amazon Basin, up to half the moisture in the air is thought to be derived from local forests. It is suspected that rainfall could decrease locally to a considerable degree when the plant cover is changed from forest to scrub or pasture land.[50]

LOSS OF SPECIES

The very complexity and diversity that make tropical forest ecosystems unique have also made them difficult to study. We still have little more than an inkling of what is required to preserve biological diversity. Many tropical forest species appear to be especially vulner-

able to habitat alteration, which makes habitat obliteration particularly devastating. Many species have specialized habitat requirements or require large patches of continuous closed-canopy forest.
The local distribution and relative rarity of most tropical species also
means that even if only a small patch of forest is destroyed many
species can be lost entirely or become locally extinct. Certain animal
species may be physically or psychologically constrained from
migrating to other patches of forest if their territory is decimated.[51]
Rarity can also result in small local population sizes and make it
more difficult to find mates. The distruption of habitats can ruin
such a delicately balanced system. If even one species is lost, others
are affected because of the interdependency of species. Coevolution
of plants and animal pollinators and dispersers in tropical forests has
resulted in many specialized relationships such that the extinction of
one species could cause the extinction of several others.[52]

Scientists can only begin to estimate how many species there are in
the tropics and extrapolate from that to estimate extinction rates.
Worldwide, around 1.4 million species of plants and animals have
been described, although estimates of the total number of species
run as high as 5 million or even to over 30 million.[53] At least half of
these, probably more, occur in tropical forests. If it is assumed that
between twenty and fifty percent of all species will be extinct by the
year 2000, we may be losing as many as six species per hour.[54]
Regardless of the actual rate of species extinctions in the tropics even
the lowest estimates are orders of magnitude higher than the
"natural background" rate of one species extinction per year over the
last 600 million years.[55]

What will result from this high rate of extinction? The short
answer is that we don't know, and won't know until it is too late.
The situation is analogous to potential changes in global climate:
it could be a major uncontrolled and irreversible experiment.
However, looking past the unpredictable future consequences of
tampering with the global ecosystem, we can speculate about more
immediate problems. Tropical forest plants and animals have
economic importance in medicinal, agricultural and industrial production. As an unknown number of potentially useful or valuable
species become extinct the economic loss could be tremendous.

Plants are particularly important to the pharmaceutical industry.
Although many drugs derived from plants are synthesized artifi-
cially, new discoveries are based upon plant secondary compounds.
In 1980, the estimated value of plant-derived prescription drugs in

the United States was $8 billion. Tropical forest plants worldwide are being screened for anti-cancer properties, and seventy percent of the plant species known to possess anti-cancer properties are from tropical moist forests.[56] Tropical forest plants may provide valuable sources of wild genetic stock for agricultural crops. Wild varieties of cultivated plants are valuable tools for plant geneticists, and varieties growing in tropical forests have already saved crops from major outbreaks of disease. For example, peanut resistance to leafspot was found in wild forms in the Amazon; its value is estimated at $500 million per year.[57] Fibers, oils and resins from tropical forest plants also figure prominently in the manufacture of furniture, clothing, varnishes, pesticides, lubricants, adhesives and many household products.

Given all the ecological constraints on the development of regions of tropical forest, is it realistic to suggest that the forests can be preserved for wildlife and put to sustainable economic use? The safest answer is that it depends on the intensity of use and on a balance between short and long term gain. It is possible to exploit tropical forests sustainably, but this usually means keeping to low levels of extraction and to long periods in which the land is left to rest and recover. Techniques which hold promise for the future are usually more expensive in the short term than methods used for the rapid extraction of forest resources. Deforestation is the result of many complex forces and of attitudes which cannot be changed overnight, but there is a growing awareness of the importance of intact tropical forest and models of sustainable development are emerging around the world.

PART II
THE CASE STUDIES

INTRODUCTION

Both the knowledge and technology exist to save the world's tropical forests. Can this knowledge and technology be employed in time? The situation of tropical forests is deteriorating, and the impact of even immediate global actions cannot be predicted. This is not to imply that there no hope left for tropical forests. But certainly, there is no room for complacency and very little time to waste. A global effort to fight deforestation will require changes both in outlook and in the economic use and control of tropical forest land.

The projects described in this book demonstrate that it is possible to establish viable reserves. Governments of tropical countries and their citizens often understand the importance of large reserves. Land can be set aside and local people can work with reserve managers to preserve forest land, and to use some sustainably. It is even possible to internalize some of the debt incurred by reserves.

Tropical forests can also be used sustainably for economic ends, although the success of these schemes depends upon long-term thinking. Most will not produce as much immediate economic gain as ecologically disastrous exploitation such as cattle ranching and uncontrolled timbering operations. If the goal changes to include both profits and the well-being of the forests, however, then ecologically-based schemes compare very favorably.

Ultimately, development efforts must shift toward the productive and economic use of cleared tropical forests. Only half of the world's tropical forest lands still support closed-canopy tropical forests and it does not seem wise to attempt risky development of the remainder. Certain fragile areas such as watersheds simply cannot support development, but in other areas, cut and regenerated land may be made productive. It is these areas, second-growth and fallow land, that should be the targets of experimental development and research into sustainable production.

The link between the disparate projects outlined in this book is more philosophical than practical. It is an attitude toward land use, a land ethic that combines an understanding of the ecological value of tropical forest with an acknowledgment that forests should be used for peoples' benefit. This view is a part of the culture of many tropical forest peoples, and is the object of the educational component of many other successful approaches. Conventional economic theory may not provide sufficient impetus for tropical forest conservation because it cannot measure the value of the intangible or unknown benefits of natural resources, but this approach, along with that of the people in cities around the world who determine tropical land use policy, will have to change. Finally, and fundamentally, these changes in attitude must be accompanied by changes in the economic control of tropical forest land. Problems of land distribution and tenure that lead to environmental degradation will only be exacerbated by the population growth in tropical areas and by the continuing concentration of wealth and land.

Another common characteristic of the successful projects is the presence of local, charismatic leaders. One or a few hardworking individuals are responsible for the establishment and conduct of new projects. This may lead to success on a local level, but can a massive problem such as tropical deforestation, so global in nature, be addressed on a small scale? We firmly believe that it can. Most of the projects presented in this book are on the grassroots level. This is not because we looked for small-scale projects, nor because we believed that large undertakings are destined to failure, but because these were the approaches that seemed to be working. Successful grassroots projects provide models that are an essential first step toward major policy changes. Concepts of forest protection, restoration, soil management and regional land-use planning must be demonstrated to work at a local level before they can be adopted as policy by national and international organizations.

Finally, the small amount of hope inspired by local success might make the possibility of any global solution more tangible. All too often the bleak scenario presented about tropical forests makes the situation appear so hopeless that the only response can be apathy. If we turn out attention to possible answers to the problem, without forgetting the grim realities, it will keep us working toward better solutions.

Chapter One
FOREST RESERVES

A first and natural reaction to the destruction of tropical forests is to wonder if the answer lies in large and strictly enforced reserves. In an ideal world enormous tracts of forest would be set aside to help preserve biological diversity and maintain the ecological "services" produced by intact tropical forests. Upon deeper examination, however, it becomes apparent that erecting a large fence around most tropical forests is not only impractical, but probably ineffective and unnecessary. Large, strictly protected reserves are surely needed to save the world's tropical forests. However, although some areas are best left in wilderness, most reserves must be carefully woven into the overall pattern of rural life. Some economic uses of tropical forest are less destructive than others, and as the projects described in this chapter demonstrate, certain economic uses may even help to preserve forests.

WHAT IS THE FUNCTION OF A RESERVE?

Closed-canopy tropical forests are the home of many species of plants and animals. In fact, biological diversity is so great that scientists can only estimate the number of species native to tropical forests. Somewhere between 2.5 million and 20 million species are squeezed into around seven percent of the earth's land surface. Preservation of this diversity should be of utmost priority. Various ethical, philosophical and economic arguments have been made for species preservation but the fact is that we simply don't know what we might be losing. Potential drugs, plants with important industrial applications, protein sources, or fascinating biological phenomena are daily being lost forever.

There is also little concurrence on the exact nature of ecological services that an intact tropical forest ecosystem might provide. But it is unequivocal that these tropical forests act to protect the watersheds of the world's great tropical rivers. The presence of tropical forest vegetation can also moderate water runoff, affecting irrigation schemes and fishing industries. Cover provided by the vegetation prevents soil erosion. Erosion can result in local devastation and also cause problems due to siltation downstream. In the Amazon Basin, topography, wind patterns and evapo-transpiration from tropical forest contribute to moisture in the air and hence rainfall.

Remnant tropical forests also provide the only truly safe seed bank to preserve biological diversity for future regrowth. Proximity to tropical forest enhances and may even be required for the regeneration of cleared land. Not only do intact tropical forests provide seed for regeneration but they harbor the animals who carry pollen and disperse seeds. Without seed dispersers, mostly wind-dispersed plants would migrate to cleared areas, and the establishment of animal-dispersed species, working in from the periphery, would be infinitesimally slow. The ties are even closer because intact forests offer food and areas for reproduction to seed dispersers. Many tropical forest animal species will not travel large distances through cleared areas.

Tropical forest reserves help protect cultural diversity as well. For thousands of years, tropical forests have been home to many indigenous peoples. Reserves can preserve the traditional lands of tribal people and protect them from incursion by developers and settlers. The forests are necessary for the physical and spiritual survival of many groups.

We have stressed the need for both basic and applied research on tropical forest ecology. Reserves are the laboratory for ecologists hoping to understand tropical systems. Although most reserves sponsor and encourage research, a few reserves have been established primarily for long-term studies, and it is at these sites that work can progress beyond surveys to interdisciplinary research. Barro Colorado Island is a small but well protected patch of tropical forest in the large lake that forms much of the Panama Canal. Originally established as an undisturbed site to study insect disease vectors, it has been administered as a general research station by the Smithsonian since 1947 and its current research is, therefore, based on data gathered for over sixty years. The Organization for Tropical Studies has encouraged research at a variety of sites in Costa Rica and has administered a small tract of lowland forest known as La Selva since 1968. In Peninsular Malaysia Pasoh Forestry Research

Center has been the focus of considerable work on Asian dipterocarp forests since the early 1970s. In addition to these classic sites are a large number of field stations scattered throughout the tropics.

Finally, reserves can offer an economic return. Sustainable agriculture and forestry, as we discuss later, are potentially economic uses of the forest. Another is the gathering of "minor" forest products including food, fibers, medicinal plants, extractive products such as latex, gums and resins, dyes, and a host of other miscellaneous products. These forest products are actually worth millions of dollars each year, and in many areas the income generated from the extractive use of tropical forests is greater than that from farming or cattle ranching on former forest land.

RESERVE SIZE AND PLACEMENT

There is no predictive science that can determine optimal reserve size and placement. Some of the ecological effects of variation in refuge size and isolation are being studied in field projects, the most ambitious of which is being conducted by the World Wildlife Fund-US and the Brazilian government near Manaus, Brazil. Of course, many non-biological factors determine the configuration of reserves, including the protection of tribal lands and the proximity to, or isolation from urban centers and development.

To ensure preservation of all species of plants and animals, all reserves should probably be as large as possible. This is particularly true in the tropics where large predatory animals, birds and insects with complex migrations, and rare species all require huge areas to ensure their survival. Although large continuous tracts of preserved intact forest are needed to maintain biological diversity, a landscape developed in an ecologically sound manner should have a range of reserves, from hedgerows and shelter belts to major parks. In light of the thoroughness of forest clearing, in most tropical areas it is likely that no reserve is too small to perform some useful function. Currently, most of the emphasis of wildlife reserves is on vertebrates, especially large mammals and birds—and for good reason: a reserve adequate for a top predator probably shelters thousands of species of smaller animals and plants. However, there are also arguments for establishing smaller reserves, including shelter belts and even living fences because invertebrates and plants may flourish in smaller areas, and these can also provide temporary refuge for larger or migratory animals. Strip parks connecting reserves could be extremely valuable in providing movement corridors for larger or season-ally migratory animals. Finally, in areas experiencing an extremely rapid rate of deforestation it might be efficacious to work on a local level to preserve many small areas, with the eventual hope of linking them.

WHY DO PARKS AND RESERVES FAIL?

Far more tropical forest land is preserved on maps than on the ground. Found around the world, these "paper parks" result from poor planning, lack of commitment, or mismanagement. Although it is usually the result of years of planning, proper reserve establish-ment is only a start; maintenance and wise long-term management of tropical forest reserves are equally important. Similarly, reserves may have unqualified government support, ample start-up funds and plenty of park guards but these are no guarantee of success. The best guarantee a reserve can have is the development of local political support in conjunction with long-term commitment and funding.

Even then, a reserve can fail. A reserve that relies entirely on government funding can find that its support has disappeared if the government changes or the country hits hard economic times. As the frontier of development marches inexorably through tropical forests every reserve will come face to face with speculators and hungry settlers who need land for subsistence farming; until a road reaches its border, an isolated reserve may need very little protection. Mismanagement has also played a destructive role: it permits ecolo-gically disastrous tourism, destructive practices to pass as sustain-able use and corrupt officials to siphon off funds or grant timber concessions.

MULTIPLE-USE RESERVES

The UNESCO Man and Biosphere concept of reserves has served as a model for multiple-use reserves, and it deserves special attention. The concept is simple: people are made part of the park. Each reserve has a strictly preserved area, usually at the center, with areas of increasing use working out toward the boundary. The edge or buffer zone is designed to be used sustainably, and people living there are considered the first defense against exploitative intruders. Many multiple-use reserves are designed as a series of concentric rings with the most strict reserve areas in the least-vulnerable center.

Although no single park features all of the zones described below, many incorporate at least three. A strict reserve that allows no human traffic may contain islands that serve as centers for scientific research, education or nature tourism. Indigenous people living in low densities practicing low-impact use of tropical forest would live in the next zone, and might serve as park guards or guides. Sustainable forestry practices and extractive uses of tropical forest would be allowed in the next zone to provide food and other necessities for people living in the park, as well as a source of revenue. Sustainable agriculture and forestry act as a buffer zone for the park. Small numbers of people could live in these areas and make a living without exploiting the strict reserve. The biosphere reserve concept makes local people the stewards of the reserve. Importantly, it also provides a method of generating income; any measure of financial self-sufficiency will help protect a reserve through possible future economic storms.

People are planning and setting up biosphere reserves throughout the tropics. Here we feature several reserves that have made considerable progress toward long-term maintenance. Several others, not included as case studies, also have promising management plans: Nilgiri Biosphere Reserve in southern India, Poço das Antas Reserve on the Atlantic coast of Brazil and Dzanga-Sangha Reserve in the Central African Republic.

ELEMENTS OF SUCCESS

Three major steps should be taken to set up a promising reserve: careful planning, establishment and provisions for long-term maintenance and support. There are no milestones by which to gauge success because a truly successful reserve should last for ever. For this reason, the best we can do here is to feature promising reserves that appear to have many contingencies covered.

ESTABLISHMENT

The aim of the reserves discussed in this chapter is to benefit people as well as wildlife. Local people participated in their creation, planning and management. Two of the reserves, the Community Baboon Sanctuary and Kuna Yala Comarca, were started by the people living in and around the reserve, and in the case of the Community Baboon Sanctuary, land was donated. Advisory councils composed of local people make policy for the Baboon Sanctuary, Kuna Yala Comarca and the Sian Ka'an Reserve. Three reserves, Cuyabeno, La Amistad and Colombia-Ecuador Binational project, were designed specifically to help tribal people living in the reserve. Benefits for people living in the periphery or outside the reserve were planned for Khao Yai, Sian Ka'an, Gandoca/Manzanillo, Dumoga Bone, Mt Cyclops, Korup and Manu.

Table 3. Qualities of Promising Reserves

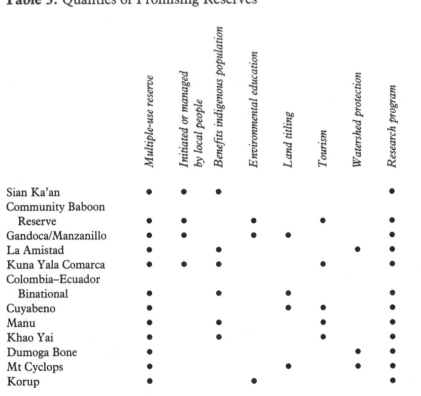

	Multiple-use reserve	Initiated or managed by local people	Benefits indigenous population	Environmental education	Land titling	Tourism	Watershed protection	Research program
Sian Ka'an	●	●	●					●
Community Baboon Reserve	●	●		●		●		●
Gandoca/Manzanillo	●	●		●	●			●
La Amistad	●		●				●	●
Kuna Yala Comarca	●	●	●			●		●
Colombia–Ecuador Binational	●		●		●			●
Cuyabeno	●				●	●		●
Manu	●		●			●		●
Khao Yai	●		●			●		●
Dumoga Bone	●						●	●
Mt Cyclops	●				●		●	●
Korup	●			●				●

Local benefits range from environmental education to land titles, community services and income from tourism. Environmental education is an integral part of the Community Baboon Sanctuary and Korup. Khao Yai, Sian Ka'an, Gandoca/Manzanillo and Korup offer extension services to teach or improve methods for sustainable agriculture. Land titling is the most important gain for people in and around Mt Cyclops, La Gandoca, Cuyabeno and the Colombia-Ecuador Binational reserves. Titling is part of regional land-use plans for these areas, which encourages local participation in developing a rational and socially acceptable plan.

This spillover of one benefit to another is no accident. Multiple-use was built into the reserves' structure and operation, and the results are impressive. The Dumoga Bone reserve, for example, has a clear role in watershed management, which helps villages in neighboring valleys which depend upon irrigation to grow rice. Sustainable agriculture is encouraged around Sian Ka'an, Gandoca/Manzanillo, Khao Yai, Mt Cyclops and Korup. Animals are also being reared for food around Sian Ka'an, Gandoca/Manzanillo and Korup. Indigenous people are living in a traditional manner in La Amistad, Cuyabeno and the Colombia-Ecuador Binational reserves. Tourism and scientific research are encouraged in the Community Baboon Sanctuary, Kuna Yala Comarca, Khao Yai, Manu and Cuyabeno.

A key addition to the reserves are their various "Friends of . . ." organizations and private support groups. The presence of these groups encourages long-term and often international interest in the reserve, and can provide continuity through changes in political climate. Private support groups provide training, supplies and other supplemental financing. They can also foster community interest in a reserve by encouraging local peoples' involvement.

PROVISIONS FOR FUTURE MAINTENANCE

The single most important factor in ensuring a reserve's ecological survival is a rational land-use plan. A good plan allows people to make a living near the strict reserve without exploiting it. Each of the reserves in this chapter has a different variation on the multiple-use plan and most have some provision for internalizing the reserve's debt. Both sustainable agriculture and aquaculture are part of the plans for the economic development zone of Sian Ka'an. Continuing environmental education efforts, hiring a local manager and possible income from tourism may sustain the Community Baboon Reserve. At Gandoca/Manzanillo, the focus is on education, the acquisition of land titles and improving the quality of life for local people. The Kuna Yala Comarca's strength lies in the Kuna's political unity and their effectiveness in garnering outside support. Its tourism and scientific research facilities could bring in income. Similarly, the program to bolster the Awas' political effectiveness will help them protect their newly-titled land. In Cuyabeno and Manu, tourism and wildlife management programs will generate income for colonists and tribal people. The tie between community social services and environmentally sound agricultural practices, in addition to possible income from tourism, will lend support to Khao Yai. The irrigation projects near Dumoga Bone should benefit it. Community involvement in the twenty-five-year land-use agreement will be advantageous to Mt Cyclops. Emphasis on sustainable development around the strict reserve, education and agricultural extension services should help protect Korup.

INTERNATIONAL SUPPORT

Reserves and parks require considerable financial support. It is probably a result of this financial burden that many reserves fail to perform even their primary function of protecting forests from clearing and development. In addition to the direct costs of operating a park there are the opportunity costs that come from removing land from immediate economic use. Many governments of tropical forest countries are not in a position to incur more debt for expensive conservation programs.

The financial problems are made even more troublesome because the support required is not a single cost, but must be maintained indefinitely. Parks can only last a short while on seed grants before an ongoing source of funding must be established. To solve the funding problems and maintain a flourishing system of tropical forest parks, it is clear that innovative approaches will need to be developed.

Many of the projects discussed in this section have focused on how to integrate reserves into the plan for regional development, often by promoting forest-based industries in the buffer zone of the reserve. However, these efforts may not be enough to generate operating funds for the park, particularly in the early years when projects are being established and trees, crops and animals are still growing to marketable size. Because many of the benefits that accrue for good tropical conservation practices accrue to us all, it would seem only just that people from wealthier, industrialized nations help support local conservation efforts in tropical countries.

However necessary this support may be, the topic of international assistance needs to be approached cautiously. In this post-colonial world, it is no longer possible, nor in our view desirable, for foreigners to dictate land-use policies to people in sovereign tropical nations. Yet many of the practices most destructive to tropical forests are carried out in order to produce commodities for the industrialized countries. Most of the large development and road-building programs have received substantial support from international and bilateral financial and assistance agencies. Furthermore, the indebtedness of the governments of many tropical forest nations to industrialized nations results in pressures for overexploitation of natural resources. In light of this, it would be irresponsible for North American and European conservation organizations not to encourage and support local environmental movements in tropical countries.

Conservationists have become increasingly sophisticated in their approach to global problems. Within the past decade we have seen the advent and development of several funding innovations, including debt exchange, ecological tourism, sister sponsorship and proposals for international conservation funds. In addition, aggressive use of the legislative process has created a specific mandate for the US Agency for International Development and other bilateral organizations to develop programs for the international protection of biological diversity.

SOURCES OF INTERNATIONAL SUPPORT

SISTER SPONSORSHIP

Sister organization sponsorship programs provide one of the most personal forms of aid between people in the temperate and tropical zones. This approach is being developed extensively by Audubon groups in the US and a number of chapters have begun to support conservation efforts in Latin America and Southeast Asia.

The goals of the sponsorship are usually specific, focused and attainable. The level of assistance varies depending on the size and wealth of a chapter, but what may seem like a small contribution can be of tremendous value to conservation organizations in the developing world. Groups in many countries, for example, cannot buy books and magazines because they lack dollars, or because foreign exchange rates are unfavorable. The Juniata Valley Audubon

Society of Tyrone, Pennsylvania, has been sending nature magazines to Peace Corps Volunteers for distribution in Belize and Guatemala, and raised $200 from (among other things) bake sales for ECODESC, an environmental group in Peru working in the Central Selva (Palcazu) area to encourage sound agricultural practices and park establishment. A ten-dollar contribution to another local fund buys copies of a Samoan bird book for distribution to school children, and helps support the production and duplication of a slide show on tropical forests to be shown in Guatemala.

At the other end of the scale, Audubon Alliance, a consortium of Audubon chapters and organizations, has pledged $12,000 annually to the Belize Audubon Society for its conservation efforts in that small but extensively forested country in northern Central America. This aid is particularly critical because the government of Belize is only recently independent, and although it is willing to set aside land for reserves, it is hesitant to incur debt by providing a large amount of financial support for park protection. The Belize Audubon Society has been given legal authority to manage these parks; international support will enable them to do so.

Perhaps the largest cooperative funding program to date has been the successful effort to purchase La Zona Protectora, an ecologically key strip of forested land that runs from the La Selva Reserve in the Caribbean lowlands of Costa Rica to the nearby mountains. This land conserves a unique altitudinal transect of tropical habitats and provides a corridor for the many insects and birds that migrate up and down the mountain. Land in Costa Rica is relatively expensive, and the purchase of 7,700 hectares required over two million dollars. The funds were raised by several Costa Rican organizations, including the Costa Rican National Parks Foundation and the National Park Service of Costa Rica, in cooperation with a consortium of major US conservation organizations including The Nature Conservancy, World Wildlife Fund-US, and the Organization for Tropical Studies. The MacArthur Foundation provided a matching grant.

NATURE-BASED TOURISM

Through nature-based tourism, travellers from developed countries can help support conservation of the spectacular flora and fauna of tropical forests. Most of the reserve projects discussed here are at least attempting to develop small-scale nature-oriented tourist facilities. The development of tourism is one of the major economic incentives for the voluntary participation of landowners in the Community Baboon project of Belize. A nature tour group has contributed extensively to the development of facilities at the Manu National Park, and it is hoped that its international prominence as a prime example of an intact tropical ecosystem will help protect it from the rapidly approaching development frontier. Cuyabeno and Khao Yai have been organized around providing wilderness experiences for the more intrepid travellers. Several of the reserves, including Mt Oku, Dumoga Bone, and Sian Ka'an, are in regions that boast unique species of birds, which will make them particularly attractive to bird watchers.

The arguments in favor of developing nature-based tourism are compelling. For one, it brings in crucial foreign exchange. In Costa Rica, Ecuador, the Philippines and Thailand, tourism ranks among the top five industries and brings in more foreign currency than wood exports. Secondly, from a conservation perspective, nature-based tourism places a monetary value on habitat preservation—a spur to governments, the local community and tourists alike to recognize that any disturbance to the environment is detrimental. Many nature travellers are willing to pay fees to help preserve the areas to which they travel. And as travellers move through remote areas of a country, the money they spend is spread throughout many rural areas that traditional tourists avoid.

Although full-scale encouragement of nature-based tourism sounds like the ideal approach to tropical forest conservation, a number of potential dangers should be remembered. Once tourism is permitted, it is difficult to control. There is often a thin line between tourism that protects the local landscape, and development that exploits and detracts from it. As the popularity of tropical forest travel increases, enthusiasts will look for more exotic and less disturbed settings. Finally, tourism, as with any luxury, is greatly affected by the world economy and political situation, and therefore cannot be considered a steady means of funding.

DEBT EXCHANGE

Although conservationists have long recognized the link between foreign debt and tropical deforestation, until recently debt forgiveness was not seen as a potential solution. Foreign debt can be purchased at discounts of fifty to ninety percent on a world market; commercial companies have been purchasing debt for several years.

Generally, debt is purchased in exchange for other equity, usually funds in local currency from the debtor government. Two recent debt-equity exchanges by conservation groups demonstrate its potential.

Conservation International helped to negotiate the purchase of $650,000 worth of Bolivian debt for $100,000. In exchange, the Bolivian government committed land and maintenance funds to expand the Rio Beni reserve in the northwestern part of the country, providing $250,000 toward maintenance and 1.5 million hectares of land. The annual operating costs of the park will be borne by the Bolivian government and conservation groups.

World Wildlife Fund recently negotiated the purchase of $10 million worth of Ecuadorian debt with Fundacion Natura, an Ecuadorian conservation organization. The funds will be converted into local currency bonds and Fundacion Natura, along with the government of Ecuador, will use the proceeds to support conservation and environmental activities, including park management and environmental education.

BLOCKED FUNDS

Another source of foreign funds of potential use for conservation are blocked funds owed to US corporations. A company's foreign assets become blocked when the government of a country is unable to honor its obligations in hard currency such as US dollars. Usually the blocked assets are money a government owes the company in payment for their products; as local currency, this money has little or no foreign exchange value. Corporations that no longer operate within the country usually carry the blocked assets on their books for years before they receive any payments; often the blocked funds are considered uncollectable. If corporations donate the blocked assets to charity they can potentially receive a quicker and larger "payoff" through tax write-offs than if they waited for a direct payment. The recipient could then use blocked funds within the foreign country to support its programs.

GLOBAL FUNDING PROGRAMS

It has been argued that tropical deforestation is too imminent for conservationists to be able to rely upon the slow and painstaking grassroots approach to fund raising. In this emergency situation, the aim should be to devise a global strategy for funding and maintaining such a system. Large-scale "action plans" that recommend investment in forest conservation have been put forward by the World Resources Institute and the United Nations Food and Agriculture Organization, and proposals have been floated for the founding of international conservation banks similar in scope to the World Bank but focused on environmental problems. Here we discuss one such global plan formulated by Ira Rubinoff, director of the Smithsonian Tropical Research Institute.

The core of the Rubinoff proposal is that a system of tropical moist forest reserves be financed from a fund made up of contributions from temperate zone nations. The economic rationale for the fund is that the destruction of tropical forests is a problem of long-term severity and will affect the well-being of temperate-zone countries. Host tropical forest countries would receive a payment from this fund to act as stewards of the tropical forest reserves. Payments would be adjusted downward if incursions into the reserve were allowed. Although full payment would be predicated on the protection of forest reserves, the actual funds could be used for a variety of programs related to conservation beyond park infrastructure, including programs for developing intensive and sustainable agriculture and forest plantations on deforested lands. Rubinoff has set an arbitrary goal of 100,000,000 hectares of forest, or slightly over ten percent of the remaining tropical forest. This is seen as an emergency reserve—the minimum required to ensure that an adequate stock of tropical organisms will survive. Currently, parks and reserves of varying quality make up only a fraction of this amount.

To fund this program, a "tax" would be levied on temperate-zone nations, scaled progressively with their gross national product. Donors would include industrialized countries, OPEC nations, the "middle-income" countries and centrally-planned economies. They could raise as much as three billion dollars a year and would be administered by a central organisation.

This and similar large-scale proposals (the World Resources Institute and FAO call for over five billion dollars) may seem unrealistic during this period of general cutbacks on foreign aid programs. The global ramifications of complete deforestation, including likely social upheaval, however, make it imperative that these proposals are considered seriously. Whatever the cost, a system such as this may be necessary if tropical conservation efforts are to succeed.

The Sian Ka'an Biosphere Reserve: Conservation of Forest and Sea, Mexico

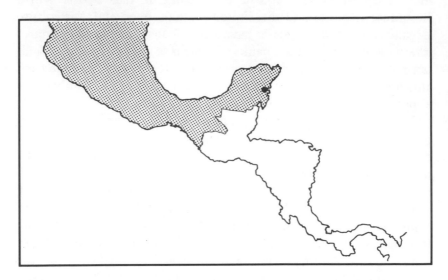

Biosphere reserves combine the protection of habitat with the preservation of traditional knowledge and the development of ecologically sound economic management. All of these activities are considered in planning a reserve, which is divided into zones reflecting its various uses. There are over two hundred biosphere reserves in various stages of planning and development throughout the world, but biosphere reserves in lowland tropical areas are scarce. The Sian Ka'an Reserve, decreed by the president of Mexico in 1986, is one of these few.

Sian Ka'an is located in the state of Quintana Roo on the southern, humid, coast of Mexico's Yucatan Peninsula. It encompasses 528,000 hectares, of which roughly one-third is covered by semi-deciduous to sub-evergreen tropical forests receiving 1200 to 1500 millimeters of rain a year. The rest of the reserve is covered by coastal swamps, savannas, lagoons and coral reef. The reserve will protect over 1,200 species of plants, 320 species of birds and many rare species of wildlife such as the jaguar, tapir, white-lipped peccary, jabiru, flamingo and ocellated turkey. Its 120,000 hectares of coastal lagoons and shallow bays form the greatest breeding grounds for the spiny lobster on the Caribbean coast of Mexico. Although not directly related to tropical forest conservation, the economic importance of the lobsters and the large coral reef (part of the second longest barrier reef in the world) provided added incentive to establish the reserve.

This part of the Yucatan has been connected with the rest of Mexico by highway only since 1960, and is still sparsely settled. Tourist development, which is spreading rapidly down the coast of Quintana Roo, has yet to reach the reserve area. These factors, plus the fact that 99.8 percent of the land was government, rather than privately owned, made the decision about where to cite the reserve fairly obvious.

Nearly eight hundred people live inside the reserve. They participate in what is termed "low-scale rational use of the natural resources," including lobster fishing in conjunction with research on sustainable yields, and intensive agriculture. Farming is restricted to one area of the buffer zone, and families are permitted to cultivate only 4.5 hectares rather than the customary 120 hectares. A model farm has been established to teach local farmers to farm intensively. Other projects include small-scale attempts at butterfly farming, the development of a management plan for exploiting useful palms in the reserve and, in the future, small tourist facilities. Residents of the reserve are allowed to gather food and materials from the forest in the traditional Mayan manner for their own consumption, but this practice, like others, is restricted to zones where wildlife will not be harmed.

The reserve involves a close working relationship between government agencies and private groups. SEDUE (Secretaria de Desarrollo Urbano y Ecologia) and CIQRO (Centro de Investigaciones de Quintana Roo) conduct several major research projects in the area. Importantly, the reserve has made considerable effort to involve people in the area. A council of representatives, composed of fishermen, campesinos, coconut growers and "Amigos", meets every two to three months.

"Amigos" are members of the non-profit private support group known as Amigos de Sian Ka'an, founded in 1986 to complement the work of the government organizations. This is a crucial role: although the reserve was established by presidential decree and the management plan provides a legal basis for its operation; without strong local support any park project in Mexico is unlikely to succeed. Amigos' main objective is the development of practical conservation projects for the buffer zone, the enhancement of local public awareness, and the generation of local and international support. Ultimately, the Amigos would like to see the reserve become economically self-sufficient. Much of the actual running of the reserve, such as protection and maintenance, will be handled by government organizations.

Sponsoring organizations: SEDUE, CIQRO and Amigos de Sian Ka'an. Funding has been provided by the Nature Conservancy and World Wildlife Fund-US.

SOURCE: Arturo Lopez-Ornat, Amigos de Sian Ka'an, A.C., P.O. Box 770, Cancun, Quintana Roo, Mexico, 77500.

The Community Baboon Sanctuary: An Approach to the Conservation of Private Lands, Belize

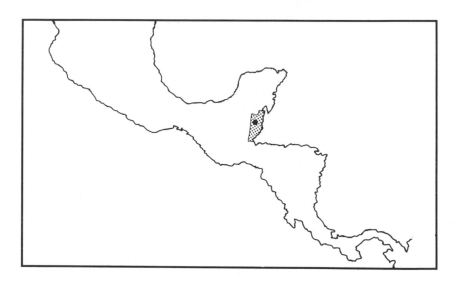

The Community Baboon Sanctuary on the Belize River is a unique reserve created with the help of local farmers and landowners. Their cooperation is essential to the reserve's success, so a major thrust of its work is education. The project must demonstrate that it meets the needs of the landowners as well as those of wildlife.

The original purpose of the reserve was to protect the habitat of the black howler monkey (*Alouatta pigra*), also known locally as the baboon. Black howler monkeys are moderate in size (they weigh about nine kilograms) and live in the canopy of tropical forests in groups of two to ten, eating leaves and fruit. Howlers are so named for the loud lion-like roaring they often perform in antiphonal group choruses. Their range is very limited, and their future survival depends upon the protection of forest within the areas of Belize, Mexico and Guatemala where they currently live. The Creole people of Belize do not hunt howler monkeys for food, a fact which facilitates the development of a community reserve project.

Called baboons in Belize, black howler monkeys (*Alouatta pigra*) and their forest habitat are protected in the Community Baboon Reserve.

Because the sanctuary was planned in consultation with people living in the area, the reserve's aims have been broadened to both protect the howlers' habitat and promote sound agricultural practices. Private landowners have agreed to use their land in accordance with reserve standards; if they do so, river bank erosion will be stemmed and the fallow time necessary for the adequate recovery of soil nutrients between clearing for traditional slash-and-burn farming will be reduced. Achieving these dual aims was a lengthy, but rewarding, process:

Site selection. The reserve site is known to be an area of high howler density and is continuous with other areas of forest habitat. Howler monkeys have few specific habitat needs beyond requiring large enough tracts of continuous forest, so their conservation is synonymous with general forest conservation.

Contact local people. The project was introduced to the community informally through discussions with landowners, village headmen, teachers and other influential people. Only after gaining their approval was an attempt made to obtain formal written permission to begin the sanctuary. During this phase, rather than present specific plans for the project, the organizers attempted to shape the project to fill perceived needs. One request made by the local villagers was that tourism be developed.

Formalization of the plan. A considerable amount of one-to-one discussion of the project took place before the formal plan was submitted. This included giving school talks and distributing a booklet, *Baboons of Belize*, which stressed their uniqueness, vulnerability, and importance as a symbol of the natural heritage of Belize, and discussed ways of helping the howlers to survive. A petition was circulated locally by a sympathetic villager to gain government support for the project. The plan was then presented at a formal meeting in a nearby town, where the development of tourist facilities was approved.

Development of the plan. Protection began at a core area which was carefully mapped with respect to the natural environment as well as to existing land tenure. A land-management plan was developed for each holding, featuring simple but important goals such as protecting a strip of forest along the river and ownership boundaries, and leaving uncut specific food trees that are known to be good for monkeys as well as domestic animals. Exactly how far the landowners would go in keeping land out of production was subject to negotiation. After a mutually satisfactory plan was worked out, the landowner was asked to sign a voluntary pledge to abide by it. In addition, the landowners received a copy of a map of their plot, a certificate from the local Audubon Society and a T-shirt. Publicity was crucial. A show on Radio Belize sponsored by the Belize Audubon Society, for example, was particularly important in mobilizing community involvement.

Publicity. Publicity not only increased local interest and pride, but helped to foster tourism, one of the goals of the local participants. Local radio and television programs, talks to conservation groups and classes, and discussions with foreign tourist groups and local tour agencies spurred wider interest. To accommodate the new visitors, local people acted as guides and families established bed-and-breakfast inns. General operating rules and prices were established and local ferrymen became informal sources of tourist information.

Expanding from the core area. The sanctuary has been slowly expanding from the core. It now comprises an area of seven villages located along 32 kilometers of riverine forest, a 0.8 kilometer strip on either side of the river and a total of 47 square kilometers of land. Seventy landowners are currently enrolled in the project. In the future, the sanctuary hopes to link up with two other wildland areas, Crooked Tree Waterbird Sanctuary and Mussel Creek Reserve.

Local management and sustainability. Lastly, a system of administration has been established. A detailed operations manual has been written, and a local manager hired. His job includes maintaining the landowners' interest in the management plans, overseeing village education about the reserve, and continuing research programs on forest phenology, soils and monkey populations. A trust fund will eventually pay the manager's salary and expenses.

Project participants: Jon Lyon, Susan O'Connell (Peace Corps), Ed Johnson, Dail Murray, Jevra Brown and villagers Fallet Young and Clifton Young.

Sponsoring organizations: World Wildlife Fund-US provided financial support and the Belize Audubon Society, particularly Jim and Lydia Waight and Mike Craig, provided logistical assistance.

SOURCE: Robert Horwich, R.D. 1, Box 96, Gays Mills, Wisconsin 54631.

Land Titling and Forest Protection around the Gandoca/Manzanillo Wildlife Refuge, Costa Rica

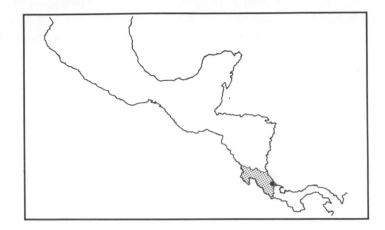

The basic premise of the ANAI land titling project around the Gandoca/Manzanillo Refuge in Costa Rica is that landlessness and deforestation go hand in hand: the greater the economic insecurity of the campesinos, the more they are forced to rely on, and therefore damage, the forests around them. The refuge protects forest and marine habitats, but to address the underlying causes of deforestation it has also founded community nurseries for economically important hardwoods and helped farmers living in the vicinity to acquire land tenure. Economic incentives will ultimately determine how local farmers use their land. Secure land tenure, however, will encourage them to employ more ecologically sound land-management practices such as leaving greater portions of their holdings in forest.

Like other tropical forest countries, Costa Rica grants campesinos "rights of possession" rather than title to their land. Possession is determined by "improvement," defined as any modification from the natural state. The most obvious and quickly profitable modification of forested land is clearing the trees. There is no clear incentive for long-term management of the land as long as there is a chance that the right of possession can be rescinded.

For land titling to be successful three obstacles must be overcome: expense, bureaucracy, and unwieldy paper work. The financially-strapped Costa Rican government is unable to cope with the near-universal need for assistance in land titling for campesinos. Although the titling program has the support of almost all local residents, some opposition has been encountered from lumber interests, cattlemen, recent colonists, marijuana growers and leftist political groups.

ANAI, a locally-based conservation and development organization, has pursued mass titling of small farms by:

- Censusing the farmers in and around the Gandoca/Manzanillo refuge;
- Defining the boundaries of the area to be titled;
- Securing the cooperation of the government and private institutions;
- Meeting with and organizing local communities;
- Marking boundaries with small paths and flagging;
- Taking aerial photographs to help define property boundaries; and
- Preparing maps of each property to be used in the titling process. The Instituto de Desarrollo Agrario carries out the final process of awarding titles.

In addition to the land titling project ANAI has begun to identify ways that farmers can make a living without damaging the forest. A network of community nurseries will support reforestation and agroforestry programs on previously cleared land. As of 1986, twenty-five community nurseries had been set up, and 1,200 farmers had raised and planted 1.5 million perennial plants on their farms, including varieties of over a hundred species, most of which are trees. ANAI also runs a two hundred hectare farm which researches the varieties most suited to the region.

The centerpiece of the project is the Gandoca/Manzanillo National Wildlife Refuge, established in 1985. Two local wardens and an administrator have been trained and now oversee and patrol the refuge. A local refuge management committee has also been selected. Protection of the beach where sea turtles lay eggs has been a major victory. In addition trails have been cut, boundary markers set out, and a headquarters constructed. Mapping and biological inventories are under way. Refuge-based economic development projects include the development of low-key tourism and cultivation of iguanas.

Sponsoring organizations: ANAI, World Wildlife Fund-US and the Jessie Smith Noyes Foundation.

SOURCE: William McLarney, Anai, Apdo 902, Puerto Limon 7300, Costa Rica.

La Amistad Biosphere Reserve, Costa Rica

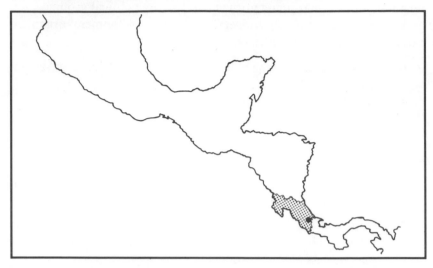

The La Amistad Biosphere Reserve project was created in the face of intense deforestation in Costa Rica and Panama. Forty to sixty thousand hectares are cleared each year in Costa Rica; little forest will be left outside of reserves by 1990. This project differs from other parks in Costa Rica, however, in its attempt to integrate traditional land-use patterns of indigenous people with strict forest protection—in other words, it follows the mandate of the biosphere reserve program. As part of an international effort, the Costa Rican park coordinates its efforts with a similar reserve in adjacent Bocas del Toro Province, Panama.

The reserve is actually a mosaic of over 500,000 hectares that include La Amistad International Park-Costa Rican sector, Chirripo National Park and Hitoy-Cerere Biological Reserve, all managed by the Costa Rican National Park Service (CRNPS). Also included are five Indian reserves: Las Tablas and Barbilla Forest Reserves (run by CRNPS and the General Forest Directorate), and the Las Cruces Botanical Gardens (run by the Organization for Tropical Studies). The reserves are contiguous and the botanical garden, located twenty-five kilometers from Las Tablas, provides classroom, dormitory and laboratory facilities.

The large size of the reserve and the wide range of forest habitats that it protects make it one of the most biologically diverse areas in Central America. With altitudes ranging from 100 to 3,819 meters and similarly extreme temperatures, the reserve contains eight out of twelve recognized life zones and protects the largest tracts of high-altitude vegetation in Central America, including oak forests, bogs and paramos. It also has good populations of quetzales, giant anteater and tapirs and has top predators such as jaguars, pumas and (perhaps) the harpy eagle.

The region's hillsides are generally steep and soils poor, so the potential for intensive agricultural development is limited. Its hydroelectric potential, however, is great and may represent one of the major future Costa Rican sources of energy.

Although only limited archeological survey work has been carried out, it is known that the area was inhabited for several millennia, first by hunter-gatherers and later, after 1500 BC, by agriculturalists. Recent research indicates that most of the reserve is included in the Gran Chiriqui archeological region, which extends from Costa Rica into western Panama. Prior to the arrival of the Spanish, Indians lived in the valleys.

Because of their isolation, the Talamanca tribes of the region were spared colonization, although expeditions and village raids took place. Even after the Costa Rican republic was formed in 1821, it took several decades before the Atlantic lowlands were firmly controlled by the central government. On the southern part of the Atlantic slope, colonization only began in the late nineteenth century, when the banana plantations were established to supply the US market. During this period an English-speaking black population settled along the Caribbean coast.

Only two of several original indigenous groups of people live in the reserve today. They live in nine communities located in five reserves on both sides of the Talamanca range. Their small numbers and shifting settlement patterns make precision difficult, but it is thought that eight to twelve thousand people live in these communities. Although their traditional way of life has changed with the introduction of new ideas and technology, and they now grow cash crops on the lower slopes of the valleys, the Indian communities have a strongly ecological approach to their environment; for them, forests are a vital resource. Parks and biological reserves are often best protected when they border on active, well-organized Indian communities.

The integrity of the reserve is threatened on several fronts. Familiarly enough for Latin America, agriculture is practiced without regard for the long-term consequences, particularly along the Pacific slope of the Talamanca Range. Forests are cleared and planted with annual crops that quickly give way to relatively unproductive cattle pasture. Compaction and burning degrade the soil and create marginal and eroding wastelands. Commercial and sporting activity continues without management or control. In the reserve itself, small coal and oil deposits have been discovered—an obvious temptation for developers. Finally, the area is ideal for growing marijuana on small isolated plots. The actual damage from this type of agriculture may be minimal, but it tends to encourage other destructive activities.

Closing the reserve off from the outside world entirely, however, would be a waste. If managed correctly, reserve development projects could benefit all of Costa Rica. Drinking water and electricity, forest products produced from areas under natural forest management, and the demonstration of land rehabilitation using tree crops are three important possibilities. Scientific and recreational tourism is a fourth. Although no facilities now exist and the area is still quite remote, a few people already visit the area for wilderness experiences.

A strategy for the conservation and development of La Amistad has been completed by a team from the Center for Tropical Agricultural Research and Training (CATIE), the National Park Service of Costa Rica, representatives of other governmental institutions and the Spanish Governmental Technical Mission (ICI). The major management problem facing the reserve is the colonization of the Pacific side. If settlers there can document their occupation of a site for ten or more years, all "improvements" must be reimbursed before they can be moved. If relocation is forceable, damages and relocation costs must also be paid. Because of these costs planners are currently determining lands that the reserve cannot afford to incorporate if it is to save ecologically and culturally important areas in a cost-effective way.

Sponsoring organizations: Center for Tropical Agricultural Research and Training (CATIE), National Park Service of Costa Rica, National Park Foundation, General Forestry Directorate and the Organization for Tropical Studies.

SOURCE: Hernan Torres, Luis Hurtado de Mendoza, and Donald Masterson of the Wildlands Program, CATIE, Tropical Agricultural Research and Training Center, Turrialba, Costa Rica.

The Kuna Yala Biosphere "Comarca": An Indigenous Application of the Conservation Concept, Panama

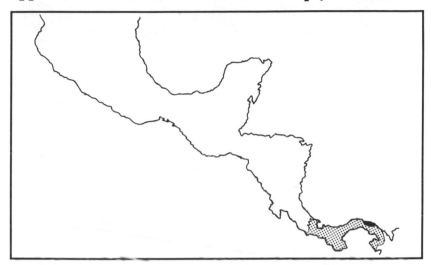

Forest conservation is an integral part of the protection of indigenous cultures in tropical areas. It is therefore no surprise that many of the recent efforts to establish innovative forest reserves have originated in native communities. Because local inhabitants of forest lands are often unfamiliar with the workings of national governments and bureaucracies, conservation projects frequently require time-consuming political organizing efforts in these communities. In the case of the Kuna Indians of eastern Panama the effort to protect pristine tropical forests from impending colonization by the Hispanic population has been made by an indigenous group that is already well organized. Although they may not have the same world view as people of other cultures in Panama, they share a sophisticated knowledge of the political and legal realities of Latin America.

The project became imperative when the first road was constructed connecting the Kuna Comarca (Comarca implies Kuna, and not Panamanian, control over land) with the rest of Panama. With the road came land speculators and colonists from other parts of Panama, bringing their practice of clearing and burning forest land every dry season. At first, the Kuna attempted an agricultural settlement on the southwest boundary of the Comarca where the encroachment was most serious. It was quickly realized that the area's steep slopes, poor soils, and high rainfall made cultivation nearly impossible. Attention was turned to a plan to develop a forest

A monkey and tapir are featured in this pattern from a mola made by the Kuna Indians of the Kuna Comarca Biosphere Reserve in Panama. Colorful cloth appliqué molas are part of the traditional dress worn by Kuna women.

reserve, now known as the Kuna Wildlands Project or PEMASKY, where medicinal plants, game, fresh water and construction materials could be extracted on a sustainable basis. A place of tremendous spiritual importance to the Kuna, the Kuna Comarca is strategically placed next to several Panamanian parks. If protected, together they will form one of the largest remaining reserves of forest in Meso-America.

Much of this project's success stems from the high degree of political and social autonomy that the Kuna enjoy within Panama. This autonomy has been won after a long history of conflict with the central government. The Kuna have full legal title to their land. Their political organization is based on participatory democracy and includes nightly meetings of local village congresses and bi-annual meetings of the traditional Kuna and general congresses.

The Kuna have participated in the project at all levels: volunteers helped set up the boundary trail system and built guard stations;

Kuna members of the Panamanian National Defense Forces patrol the access road; Kuna technicians are instrumental in planning and administering the reserve; and various Kuna organizations, particularly the Kuna Employees Association, work to gain financial, legal and political support from outside the Kuna community. Their accomplishments so far include the construction of the Nusagandi headquarters with a laboratory for scientific field work, small-scale tourist facilities (a restaurant and dormitory building) and boundary-line patrol trails to keep colonists from overflowing into the reserve.

The Kuna have enlisted expert help from a number of outside organizations, including the Tropical Agricultural Research and Training Center (CATIE), Smithsonian Tropical Research Institute (STRI), World Wildlife Fund-US, the Inter-American Foundation, USAID and various ministries of the Panamanian government. Technical advice on topics from agroforestry to construction has been provided through seminars, workshops and courses. CATIE and STRI have contributed toward a biological and resource inventory of the Comarca.

The success of the Kuna's efforts has already made it a model for other indigenous groups in the Americas. The Embera, who live in the adjacent Darien region of Panama, have begun consulting with the Kuna to develop a similar plan for their lands.

Sponsoring organizations: CATIE, World Wildlife Fund-US, USAID, the Inter-American Foundation and the MacArthur Foundation.

SOURCE: Temistocles Arias, AEK/PEMASKY, Apto. 21012, Paraiso, Panama; and Brian Houseal, The Nature Conservancy International, 1785 Massachusetts Ave. NW, Washington, DC 20036.

A Bi-National Approach to the Protection of Indian Lands, Colombia and Ecuador

Protection of the diverse indigenous cultures of the New World tropics is tantamount to conservation of the forest and river resource on which the people depend. Unlike the Kuna, most Indian groups in South America do not have a long history of political interaction with central grovernments, and so they are less able to organize

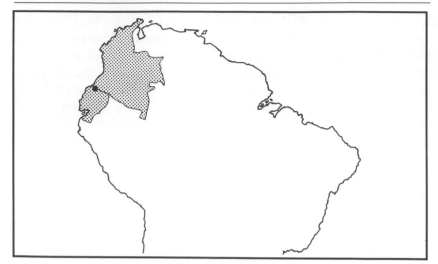

against the myriad threats of incursion onto their lands. The purpose of this project is to establish an area of the Pacific lowlands along the border between Ecuador and Colombia in Carchi and Esmeraldas provinces. This area has escaped some of the problems of rapid development plaguing other areas of Andean forests. The forest is exceedingly wet and not attractive for guerrillas or drug growers. In the late 1970s, however, rumors of a World Bank-financed road brought an increasing number of colonists from Colombia. The establishment of the reserve required a great deal of basic work in organizing and establishing the land rights of the Awa Indians.

The nucleus of the Awa demarcation project was the creation of the La Planada reserve in 1982. Covering 1,700 hectares of forest, it has helped form a nucleus for a regional resource management plan to preserve natural resources for the local Indian and Hispanic residents.

The Awa project, begun in 1984, is aimed at securing Indian land rights in the lands surrounding the La Planada reserve. Toward that end the project is registering Indian residents and delineating Indian lands in both countries. In conjunction with the national governments and private institutions, forest resources have been inventoried and a program for their effective use established.

Work has progressed the farthest in the Carchi Province of Ecuador where by 1986 about a thousand Awa had been registered. Residents are receiving training in resource management and organization and biologists from the University of Aarhus, Denmark, and the Catholic University, Quito, have completed their resource inventory. The entire project will eventually cover roughly 104,000 hectares. Research to assess the impact of the project has been contracted to the National Institute for Peasant Capacitation by the Ecuadorian Ministry of Agriculture and Livestock. Support for the efforts to demarcate and title lands to the Awa and to promote effective indigenous organization has come from Cultural Survival and the Ecuadorian government. The work was recently extended to the forests of Esmeraldas province of nearby Colombia where another thousand Awa reside.

Considerable effort has been put into training programs, particularly for developing the political and organizational skills of the Awa so that they can better deal with external agencies. Because the Awa have been isolated until relatively recently, they have little experience with central governments. The Awa themselves have lived in scattered villages with relatively little overall organization, but through grass-roots organizing even the most isolated villages have been linked through the formation of regional incorporated organizations, known as Cabildos. The local groups are then integrated into a National Coalition (CONACNIE), which has attempted to promote indigenous claims while developing sustainable land use by mestizo farmers around the La Planada reserve. Finally, it has achieved bi-national cooperation of the central governments. In 1986, the joint commission on Ecuadorian-Colombian Amazonian Cooperation officially endorsed the Awa reserve project.

Sponsoring organizations: Cultural Survival supported indigenous land claims; Foundation for Higher Education and World Wildlife Fund-US funded La Planada.

SOURCE: Richard Reed, Cultural Survival, Cambridge, Massachusetts 02138.

The Cuyabeno Wildlife Production Reserve, Ecuador

The Cuyabeno Reserve is a model modern tropical reserve. Located in an area extraordinarily rich both biologically and culturally, it stands at the frontier of Amazonian development. The reserve plan incorporates strict protection of forest and streamside habitats, preservation of a Siona-Secoya Indian reserve, and tourism and wildlife

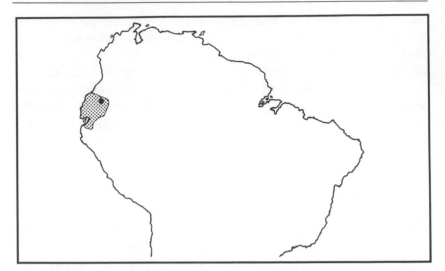

management programs. Such is its success that the Ecuadorian Department of National Parks and Wildlife ranked its master plan first among those of the nation's reserves, and has held it up as an example for other reserves to follow.

Legally created in 1979, the reserve covers 254,760 hectares of *terra firme* and *varzea* (flood plain) forests in the extreme northeastern portion of the Amazon Basin, roughly two hundred kilometers from Quito. Its selection as a reserve site after a long survey of important natural areas between 1974 and 1979 is not surprising: the land, drained by tributaries of the Amazon, is dotted with oxbow lakes and contains an unusually diverse collection of plant and animal species.

The reserve's purposes are four-fold: the protection of native vegetation and wildlife; the protection of indigenous human communities; the management of wildlife projects; and the development of tourist and education facilities.

PROTECTION OF NATIVE VEGETATION WILDLIFE

Because no one knows exactly what plants and animals live in the reserve, their protection is somewhat simplified. Particular species can be protected by targeted environmental management. But in a situation such as this, where an initial inventory will almost certainly discover new species, protection of almost all species depends upon the maintenance of large tracts of pristine habitat. Most of the habitat in the reserve is relatively undisturbed; a major exception is the westernmost sector, where colonists have settled along oil exploration roads. Land in this area will be ceded to IERAC (Institute for Colonization and Agrarian Reform) which will provide land titles to settled families, and in exchange a larger area of forest land will be added to the reserve. An additional hundred or so families that have settled in the core of the reserve will be provided land titles outside the reserve, and no further land titles in the reserve will be granted.

PROTECTION OF INDIGENOUS COMMUNITIES

One of the goals of the reserve is to help minimize the impact of development on the indigenous Siona Indians who live on the reserve and on the Sionas and Secoyas who live just beyond its southern edge. These peoples get their food by traditional means — fishing, hunting and cultivation. While they can hunt and fish freely, they are encouraged not to kill rare animals. Just before the founding of the reserve, 410 Sionas received legal title to their land. This fact, plus the existence of the biological reserve, may act as a buffer to the direct effects of encroaching colonization. From a scientific viewpoint, the Siona's knowledge of their environment will contribute to an understanding of the potential uses of tropical forest plants and animals.

WILDLIFE MANAGEMENT PROGRAMS AND TOURISM

Two further efforts are being made to promote appropriate development and to provide income to local villages. The first is the establishment of wildlife management programs that, through the rearing of native species such as the peccary and paca, will provide economic benefits to peasant communities. The second is tourism. Presently, the reserve boasts only the most rustic of facilities; adventurous travellers can be guided by Siona Indians on a canoe trip. Now, additional tourist facilities are being developed by the reserve administration in coordination with local communities and Quito-based nature tour companies.

Cooperating organizations: World Wildlife Fund-US and the Department of National Parks and Wildlife of the Republic of Ecuador.

SOURCE: James Nations, Center for Human Ecology, 2106 Rundell Place, P.O. Box 5210, Austin, Texas 78763, and Flavio Coello Hinojosa, Director, Reserva Faunistica Cuyabeno, Dept. de Parques Nacionales y Vida Silvestre, Direccion Nacional Forestal, Ministerio de Agricultura y Ganaderia, Quito, Ecuador.

Manu Biosphere Reserve, Peru

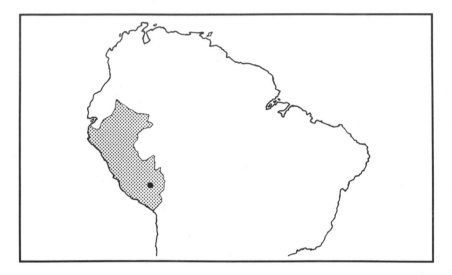

Despite the intensity of tropical deforestation, there remain large areas of relatively pristine wilderness. These scarce undisturbed ecosystems, protected primarily by their remote and inaccessible locations, are now under threat by ever-extending roads and their logical consequence, colonists. To preserve naturally functioning tropical forest communities requires the establishment of carefully guarded mega-reserves. One attempt to establish such a large strict nature reserve in the face of encroachments can be found in the Manu Biosphere Reserve located in the Amazon Basin near Cuzco, Peru.

The Manu Biosphere Reserve is the largest national park in the world; the strict reserve alone covers over 1.5 million hectares, while an additional 300,000 hectares provide a zone of sustainable development. The park protects over 1,000 species of birds, 13 species of monkeys, 110 species of bats and over 15,000 species of plants. A few square kilometers of forest can have over a thousand species of flowering plants. But the reserve's importance cannot be counted only in numbers; it also harbors a unique collection of species. It is one of the few areas in the neotropics where large predatory animals, primates and ruminates are common.

So far the success of the Manu as a preserve for wildlife has been based on its remote location, bounded to the west by steep mountains and to the east by waterfalls that prevent easy navigation by large boats. Until quite recently this region of eastern Peru has remained relatively untouched by westerners, with the exception of an occasional rubber collector. For the entire department of Madre de Dios, where the Manu is located, development has been increasing in the past thirty years as roads push over the Andes to the port cities. Throughout this period, Manu has become even more economically isolated since activity has focused on areas accessible by river and overland transportation.

Although in the 1960s loggers cut mahogany trees from the forest close to lower stretches of the Manu River, the wildlife has been subjected to minimal hunting by local people. Since the cessation of these logging activities, the wildlife of the Manu region has been virtually undisturbed. Because of the remarkably tame and abundant wildlife, the Peruvian government established the Manu National Park in 1973, which strictly protected the forest in the upper reaches of the watershed and formed a protection zone in the lower more recently logged sections. The Manu National Park has been an official part of the UNESCO Biosphere Reserve system since 1977.

Despite its recognition by the Peruvian government and the international conservation community, the development frontier is moving closer. A road is being built to connect the Manu area to the road network of neighboring western Brazil. In the face of impending colonization, there is still some time to form the economic basis for long-term maintenance of the Manu Reserve. The reserve includes a cultural zone which allows low-level exploitation of forest by some four hundred residents (around a hundred of whom are Indians) and an area for aquaculture, such as fish farming. The major forest-based economic activity that is being developed in the reserve zone is nature-based tourism, an activity which is consistent with strict preservation of wildlife.

Tourism in the Manu has been discouraged by its sheer inaccessibility; until recently it required days of travel to get into and out of the park. Recently, however, local tour companies have developed nine or ten-day trips into the area. These local companies often cooperate with nature tourism companies in the United States, which provide their own expert guides. One US company, Victor Emmanuel Tours, has provided funding for a lodge and an airstrip which places the Manu within a 45-minute plane flight from Cuzco.

Nature-based tourism increases the constituency of international support for the biosphere reserve. In addition, it provides direct support for the operation of the park through user fees and voluntary contributions from tour companies and individuals to conservation organizations. Most importantly, it provides a real source of cash to people in an area that provides few other financial opportunities. In 1985, it was estimated that between a quarter-million and a half-million dollars went into the local Peruvian economy in the form of payments to tour companies, boatmen, route guides, lodges, etc. The continuation of this money is directly dependent upon preservation of the wildlife that attracts the tourist.

Although the subsidization of tropical conservation by naturalists and scientists through tourism from around the world may not be a general solution to the problem of tropical deforestation, it may prove to be the only effective strategy for completely preserving certain model ecosystems.

Cooperating organizations: Peruvian government's Corporation for Development of Madre de Dios, General Directorate of Forestry and Fauna of the Peruvian Ministry of Agriculture, World Wildlife Fund International, Conservation Association for the Southern Rainforests of Peru (ACSS), Portal de Panes 137, Plaza de Armas, Cuzco, Peru, Peruvian Association for the Conservation of Nature (APECO), Parque Jose Acosta 187, Altos, Magdalena, Lima 17, Peru, Association for Ecology and Conservation (ECCO), Vanderghen 560-2A, Lima 17, Peru, and Peruvian Foundation for the Conservation of Nature (FCPN), Chinabon 858-D, San Isidro, Lima, Peru.

SOURCE: Charles Munn III, Wildlife Conservation International, New York Zoological Society, Bronx, New York, 10460.

Protection and Development in and about Khao Yai Park, Thailand

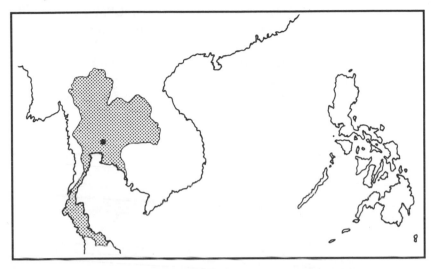

It has become increasingly apparent that national parks cannot be protected without addressing the needs of the people who live in the surrounding lands. This realization has spawned two projects dealing with Khao Yai National Park in Thailand, one concerned mostly with park management, the other emphasizing community development.

The first project, which is ultimately aimed at both elephant and forest conservation, was an ambitious drive to create an effective conservation program in and around the reserve. Supported by World Wildlife Fund/IUCN through the Royal Thai Forest Department, the project started with three major goals: to produce a management plan for the Khao Yai Park, which would be the first such plan for a protected area in Thailand, serving as a model for such plans in other parks and sanctuaries; to carry out research on the distribution, habitat use and number of elephants in different parts of the park; and, finally, to initiate a rural development program in the nearby village of Ban Sap Tai which would reduce local villagers' dependence on the park and their consequent incursions to cut trees, plant maize and poach plants and animals.

The Khao Yai management plan, the first of its kind in Southeast Asia, was developed by a team from the National Parks Division of the Royal Forest Department and other government agencies. Because of its emphasis on providing benefits to local villagers, the plan sets a precedent in Thai policy on national parks. Its country-wide management planning mechanism, incorporated in the Thailand National Economic and Social Development Plan, calls for the preparation of twenty-three plans for national parks and wildlife sanctuaries.

Wild elephants figure prominently in the Khao Yai plan for a variety of reasons. The elephants themselves attract sight-seers, and, for more active tourists, they provide an important service: once

Wild Asian elephants are protected in Khao Yai Park in Thailand. Elephant trails are also used by tourists.

mapped, the trails they create become trekking routes. A growing group of international hikers now provide a small but significant source of revenue for local villagers, some of whom act as guides.

The second project, "Rural Development for Conservation," was begun in Ban Sap Tai in 1985. Villagers are offered agricultural assistance to help them diversify their farming, some health care and other social services, help in marketing and education in conservation techniques. The project also promotes limited tourism. The Population and Community Development Association and the Wildlife Fund-Thailand run the project; Deutsche Welthungerhilfe ("Agro Action") provides the funding. Agricultural assistance will be continued in Ban Sap Tai and is now, with the help of USAID, being extended to several neighboring villages.

Credit, education and collective business ventures are the three project priorities in improving economic conditions in the villages. Local moneylenders were charging annual interest rates of over fifty percent, and farmers found themselves trapped in a seemingly endless debt spiral. In response to this problem, a community-based credit cooperative, the Environmental Protection Society (EPS), was established. Membership was extended to all villagers who possessed land titles and who promised not to break the national park regulations; in return, families could get loans at a more reasonable twelve percent. In addition to supporting promising agricultural activities, EPS also acts as an extension service and provides a family-planning service.

EPS activities in the villages have reduced hunting in the park, eliminated encroachment and encouraged the peaceful withdrawal of several families who had built homes in the park. A small private company is developing tourist facilities. Nonetheless, the work in and around Khao Yai Park is still considered a demonstration project; its rural development efforts must continue to ensure environmental success.

Sponsoring organizations: First project: World Wildlife Fund and IUCN. Second project: Deutsche Welthungerhilfe, Wildlife Fund-Thailand and the Population and Community Development Association, Thailand.

SOURCE: Warren Brockelman, Center for Wildlife Research, Faculty of Science, Mahidol University, Rama 6 Road, Bangkok 10400, Thailand; Robert J. Dobias, 124/50 Lad Prao, Soi 109,

Bangkok 10240, Thailand; Charlurmchat Khonthong, PCDA, 8 Sukhumvit 12, Bangkok 10110, Thailand; and Martha Van der Voort, World Wildlife Fund-US, 1250 24th Street NW, Washington, DC 20037.

Protecting Wildlife and Watersheds at Dumoga Bone, Indonesia

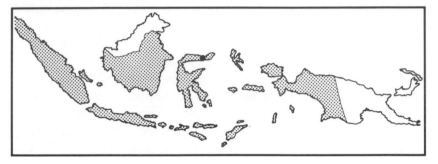

Most tropical forests grow on land that is inappropriate for intensive cultivation. Even where the soils are fertile, their productivity is often threatened by the deforestation of nearby watersheds. Torrential monsoons can be absorbed and slowly released by forest-covered land and even harnessed to irrigate water-loving crops such as rice— or, if the area is deforested, they can wash off denuded hillsides, carrying rivers of eroded mud and debris to flooded bottomlands. Agricultural conservation and development is therefore crucial to forest protection. It is precisely this reasoning that led to the founding of the Dumoga Bone Forest Reserve as part of a massive World Bank loan for irrigation projects in Indonesia.

The Dumoga Valley, nestled in the mountains of northern Sulawesi, is a fertile watershed protected until recently by its remoteness and malarial swamps. In the 1970s, a major highway system connected to the provincial capital, a government-sponsored transmigration project and the construction of a World Bank-funded irrigation system ended the valley's tranquillity. Between 1977 and 1980 its forests were cleared almost frantically in an effort to open up new agricultural land. Some trees were logged, but many were simply burned on site. Most of the forest was unprotected and open to homesteading. Even the protected forests on steep slopes were poorly guarded from the onslaught of settlement. The prospect of irrigation for rice production pushed land prices up and promoted speculation and more deforestation, since land had to be cleared to obtain legal possession. Clearing in anticipation of increased land values was practiced by the original settlers, as well as by the proxies of absentee landlords.

By the late 1970s it was clear that the entire irrigation scheme of the Dumoga Valley would be threatened with both water shortages and flooding if logging and clearing in the surrounding hills continued unabated. In 1980, the existing 93,500-hectare Dumoga and 110,000-hectare Bone Reserves were consolidated into a special Dumoga catchment project, also funded by the World Bank. The costs of the irrigation system came to about $20 million, while the protection of the catchment area cost $700,000. The World Wildlife Fund and the Indonesian Nature Conservation Department provided expert advice for the new 300,000-plus hectare national park.

The irrigation project, the demarcation of the Dumoga Bone National Park and the construction of a park headquarters were completed in 1987. Because a considerable amount of clearing had already occurred on the lower foothills, some of the catchment could not be included in the park. Additional settlement and clearing had continued several years after the park was established, so reducing the amount of water available to the irrigation project. In response, the Indonesian government began a campaign to reverse encroachment. Four hundred squatter families were evicted from settlements within the park. They and many other families were resettled, and the prevention of further colonization has been vigorously pursued. Controls on land registration near the park were tightened and boundaries have been more firmly marked by planting rows of fast-growing trees.

With the new irrigation system, the valley has changed from a rice-importing to a rice-exporting area with the biggest increases in agricultural income reaped by the farmers within the irrigation project area. Unfortunately, the original inhabitants of the valley have benefited least from the switch from dry-land to wet-land cultivation. Having sold their holdings and moved to the hills, most were unable to be part of the development programs and were more likely to be living on the reserve, and therefore to be evicted from the park or resettled.

The Dumoga Bone Reserve represents one of the largest and best enforced nature reserves in Asia. Its protected, highly endemic flora and fauna have been studied by scientists working under the auspices of "Operation Wallace," the largest biological expedition ever, organized by the Royal Entomological Society; hundreds of hitherto unknown species have been discovered. Although the reserve primarily protects highland areas in the catchment, certain key lowland habitats, harboring a diversity of species, have been included as well.

Sponsoring organizations: Funding is provided by loans from the World Bank and the Indonesian government; park planning has been conducted by World Wildlife Fund and the Indonesian Department of Nature Conservation.

SOURCE: Robert Goodland, World Bank, 1818 H Street NW, Washington, DC 20433.

A Community-Managed Buffer Zone for a Nature Reserve, Indonesia

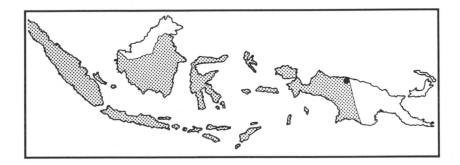

One of the problems facing Third World governments is the difficulty of regulating land use. The areas of public forest are huge, the infrastructure to patrol and protect them is weak, and the pressures to colonize and develop are tremendous. Without strong local political support for forest management, government control over public land is virtually impossible. This is true not least in Indonesia, where the Indonesian Ministry of Forestry is responsible for 143 million hectares of land, roughly three-quarters of the entire country. In 1986 alone, as many as one million hectares of forest were cleared. To address these problems the Indonesians have established a social forestry program based on ecologically sound forest management developed at the community level.

The Mt Cyclops Reserve, located in New Guinea (Irian Jaya), is one of twenty-four pilot programs in social forestry and one of the first attempts to create a community-managed buffer zone. The pilot program has been the result of a collaborative effort between the Ministry of Forestry, the provincial government of Irian Jaya, World Wildlife Fund-US, and the Ford Foundation.

The 31,600 hectare reserve protects montane tropical rainforest and the watershed for a major reservoir and hydroelectric power plant. However, the reserve has suffered encroachment from both indigenous and migrant farmers, as well as loggers, hunters and developers. Tribal groups have been increasingly pressured to sell their land rights to local or outside developers for large-scale logging and farm projects. The focus of the project, at least initially, is to stabilize land claims and to establish acceptable patterns of land use.

The project began with a meeting of a working group consisting of members of the provincial planning board, the Department of Forestry, the local university, and the Irian Jaya Development Foundation (YPMD), which is the most important non-governmental organization in the province. The group determined that the highest priority for the Mt Cyclops Reserve was the resolution of land claims between the various tribal groups and the establishment of firm boundaries for the reserve. In addition, it was decided that an effort should be made to set up a new process for developing acceptable land use practices by the local communities and governments.

The first step was the selection of a ten-person team comprising park rangers and NGO field staff. For four weeks they were trained in government land use methods, indigenous legal systems, land and agroforestry management practices and field techniques such as mapping and interviewing. The team members, assisted by staff of the YMPD, the forestry department and WWF, were then sent into areas where they were familiar with local customs and languages to discuss land use needs with tribal leaders and clan members.

One such discussion took place in the village of Ormu Besar, located on a small bay at the foot of the Mt Cyclops massif. The village is inhabited by several hundred people who use the surrounding lands for gardening, hunting and gathering, and fishing. Although there had been preliminary discussions between the park staff and the villagers, no formal agreements regarding boundaries had been reached.

At the first meetings, village leaders expressed their concern that the needs of the local people would not be met by a buffer-zone agreement. Further discussions focused on land use needs. Using sketch maps and aerial photographs, the field team helped the community to assess its current use of the land. The community decided that it wanted to maintain a strip of coastal land up to 300 meters in elevation for traditional mixed gardening and groves of the sago palm, a staple food. This proposal was agreed by the forestry team, and an additional zone of low-intensity use up to 500 meters in elevation was established where villagers could cut specific trees, hunt boar, collect fuelwood and quarry stones for ritual exchange. Also under discussion were plans for expansion, the more intensive use of cultivated land and a small visitor facility at the head of one of the wilderness trails. While other tribes were selling their land to outsiders, the Ormu people had come to a clear agreement with the government regarding their tenure status and in the process had helped establish a key environmental buffer zone.

Sometimes it is the small touches that enhance the success of a project. In this case, the field teams believe that the use of oblique aerial photographs on a scale of 1 to 8,000 was an inexpensive and effective way of stimulating discussions concerning land use needs. Photographs can be made from a light aircraft with a standard single-lens reflex camera and be processed in a basic darkroom. Then agreed upon boundaries can be placed on a formal map and incorporated into a contractual agreement. Another innovation was suggested by the Ormu leadership, who suggested that the project use the traditional marker for boundaries a shrub, *Coryline terminalis* with its bright foliage, as well as natural topgraphic features.

The current land-use agreements will run for an initial twenty-five years; if the goal of establishing agreements with four tribes is achieved, most of the area around the reserve will be under the management of these locally-based buffer-zone plans.

Sponsoring organizations: Irian Jaya Provincial Planning Board, Provincial Department of Forestry, Irian Jaya Development Foundation, World Wildlife Fund-US, the Ford Foundation and the East-West Center.

SOURCE: Mark Poffenberger, Ford Foundation, Southeast Asia Regional Office J1. Teman Kebon Sirih 1/4, Jakarta, Indonesia.

Korup National Park, Cameroon

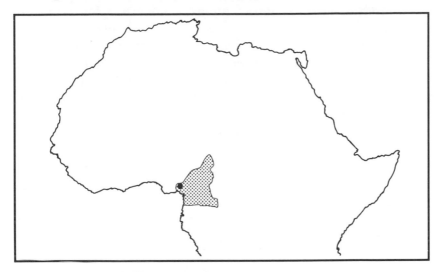

Africa's rainforests have been halved in size in the past century. The continent now contains some nineteen percent of the world's tropical rainforest. In many West African countries the forest has been extensively cleared; conservation is now essentially a salvage operation. In Cameroon, fortunately, there is a greater emphasis on habitat preservation. The government is committed to the protection of twenty percent of its territory. Most of its national parks have been located in the savanna zones of the northern and central parts of the country, leaving the southern rainforests relatively unprotected. Now, however, the government is creating three parks in the south. One of these is the Korup National Park.

Korup National Park is an integral part of a regional development program that includes the protection of an area of pristine, species-rich forest. Much of the project deals with agricultural extension services, agroforestry schemes, development of alternative sources of protein, improved strains of food and cash crops around the border of the park.

Most of Korup is located in the Ndian Division of the southwest province of Cameroon. Korup has been extensively surveyed botanically and is the most species-rich African rainforest for which comparable data exist. Largely unaffected by settlement, it is thought to be a former lowland Pleistocene refuge area and is now inhabited by many unique species. With over four hundred tree species belonging to more than fifty families, and some fifty species of mammals of twenty different families, Korup is of inestimable value to scientists and conservationists. It is exceptionally important for primates, sheltering almost a quarter of Africa's species. A total of 252 bird species have been observed in Korup and its immediate vicinity.

While Korup was established as a forest reserve in 1937, the national park was created by presidential decree in 1986. This gave it the highest possible degree of protection under the laws of Cameroon. The park covers 126,000 hectares, with almost 300,000 hectares included in the rural development project.

Korup is viewed as a model for forest parks in tropical Africa. Much of the project is still in the planning stage, but it will eventually cover every aspect of conservation of forests and resources. Its goals include:

- providing people with alternative sources of protein to replace the animals currently being taken from the park;
- producing land use and land suitability maps;
- identifying suitable sites for the resettlement of villages now in the park;
- designing sustainable farming methods for forested areas;
- building roads;
- training and educating local people and staff, particularly about environmental issues;
- developing sustainable, non-destructive uses of the forest;
- establishing a scientific research program to assist in the informed management of the park; and
- creating the infrastructure for tourism.

Korup has benefited from the fact that half of the experts involved in its creation were from Cameroon while many of the others were familiar with the area. It has gained formal recognition, raised funds and fostered cooperation between government ministries and research groups. By July 1987 a British-sponsored agroforestry program had begun and a professional park conservator was appointed to supervise a park staff of eight. Its future, however, is likely to be more difficult. A thousand people still living in the park must, under Cameroon law, be resettled. More fundamentally, Cameroon is now entering a period of economic austerity. Finance for national parks, as elsewhere in the Third World, is becoming increasingly scarce.

Sponsoring organizations: British Overseas Development Administration, World Wide Fund for Nature, National Institutes of Health, Missouri Botanical Garden, Parcs Canada, International Primatological Society, USAID.

SOURCE: Stephen Gartland, Korup National Park Project, c/o Pamol, Ndian Estate, BP 5489, Akwa, Douala, Republic of Cameroon, West Africa, or Wisconsin Regional Primate Research Center, 1220 Capital Court, Madison, Wisconsin 53715.

Chapter Two
SUSTAINABLE AGRICULTURE

CURRENT FARMING PRACTICES

SHIFTING AGRICULTURE

Swidden-fallow or shifting agriculture is the most widely practiced form of farming in the tropics. It is also the single most important cause of tropical deforestation. With as many as 240 million people growing food for subsistence in rural tropical areas, at least eight million hectares of forest are converted to cropland each year—a problem which will be compounded in coming decades by population growth and the concentration of land ownership. No single miracle crop nor farming technique can save tropical forests, but the modification and more careful practice of farming techniques might slow the ever-increasing use of tropical forest land for agriculture.

Shifting agriculture is not inherently destructive of forest land. Tropical forest peoples have engaged in a variety of shifting agriculture methods since the advent of farming. In fact, recent evidence shows that many forested areas considered "virgin" or untouched were occupied for centuries by people practicing shifting agriculture. Today there are still people practicing swidden-fallow agriculture in a sustainable fashion. The problem is not the farming technique itself, but that too many people are trying to farm too little land, and the fallow period between plantings in which the land recovers is being shortened or altogether eliminated.

In traditional swidden-fallow agriculture, a small forested plot is cleared and the slash, or cut vegetation, is left to dry. Before the slash is burned, certain root crops are planted, and these and other crops planted later are nourished for several seasons by the nutrients released into the soil by the ash left by the fire. After two to three years, the plot is unable to sustain annual crops. It is then left to fallow, and a new plot is cut. But the old plot is still useful. While the forest regenerates, perennial shrubs and trees can be harvested for fifteen to thirty years before the plot is cut and burned again. This system requires between one and five hectares per family each year in the neotropics, but if practiced carefully, no primary growth forest need be cut after fifteen to thirty years. Many groups actually cultivate seemingly abandoned fallows for many years.

Shifting agriculture ceases to be sustainable in areas of intense population pressure and where immigrant settlers are unfamiliar with local conditions. So long as some forest surrounds a farm, there is some protection against agricultural pests and weeds, and an ongoing source of nutrients. But once new settlers remove the forest buffer, weeds and pests spread, and soils become susceptible to erosion. If new land is unavailable and plots are cut after too short a fallow period, the soil becomes unable to sustain crops. Land-tenure patterns tend to exacerbate the situation because most shifting agriculturalists do not have legal rights beyond use-rights to the land they farm. They cannot protect fallow land and lack the incentive to practice soil-conservation techniques or improve land they do not own.

LARGE-SCALE AGRICULTURE

Large-scale agriculture and agribusiness occupy far less land than small farmers' plots, but they are equally destructive. This is most obvious where forests have been cleared to make way for export crops such as palm oil, rubber or citrus. For one thing, monocultures are especially vulnerable to disease and pests. Of more lasting damage is the less direct process of concentration of land in the hands of large landowners, and the repercussions this has on the environment. A wealthy minority usually controls most fertile land; the practitioners of subsistence agriculture are left to farm poorer soils and steeper slopes. Massive soil erosion and poor yields force small-scale farmers to move more often, leaving abused lands behind.

RANCHING

In some areas, particularly in the New World tropics, livestock is considered an ideal business venture. It requires minimal investment in capital and labor and is often subsidized by tax incentives. On a per-hectare basis ranching is often only marginally profitable. But when the rancher can put his cattle on land cleared and abandoned by colonists, profits instantly improve. It is not only that ranching encourages the clearing of forests; since their cattle graze over large areas, many ranchers show little regard for land conservation. Improving grazing practices and producing fodder more intensively, along with eliminating tax incentives, are essential if ranchers are to be dissuaded from clearing new forested areas for pasture.

SUSTAINABLE SYSTEMS

To be sustainable, an agricultural system must maintain crop production over time. Whether plots are permanent or shifting, they must provide adequate nutrition for crops, afford some protection against weeds and pests and be able to withstand fluctuations in environmental conditions. Social and economic conditions in subsistence agriculture may preclude a reliance upon costly or dangerous supplements such as expensive equipment, commercial fertilizers and pesticides.

ADHERENCE TO ECOLOGICAL PRINCIPLES

The ultimate aim of tropical agriculture should be to produce as much food as possible without degrading the land. Agricultural practices that contribute to tropical forest conservation are slightly more difficult to define. Many good systems involve the careful use of land so that potentially less forested land is required, but this does not necessarily contribute to forest conservation. In fact, this type of agriculture could simply result in more farmers being able to use the same amount of land. Ideally, intact forest should be part of an agricultural system, as it can slow the spread of pests and crop diseases, provide seed stock for faster forest (and nutrient) regeneration and provide supplementary food and household necessities. From a conservationist's point of view desirable agriculture requires the close proximity of untouched forest.

As the following case studies illustrate, sustainable agriculture is a fairly all-encompassing term. It includes subsistence and market-oriented farming, a variety of ways of producing animal protein, and several forms of indigenous agriculture, some traditional and some relatively new. Each of these is area-specific; that is, they result from the growing conditions, culture and social and economic structures

Table 4. Qualities of Sustainable Agriculture Approaches

	Successional	High crop diversity	Topographic considerations	Tree crops	Nutrient enhancement	Small plots	Integrates domestic animals	Cash crops
Mayan Agriculture	●	●	●	●	●	●	●	
Intensive Agriculture on Limestone					●	●		●
Beef to Dairy Cattle					●		●	●
Iguana Ranching				●			●	●
Kayapo	●	●	●	●	●	●		
Japanese Farmers at Tome Acu	●	●		●	●	●	●	●
Bora	●	●	●	●	●	●		
Peruvian Agroforestry	●	●		●	●	●		●
Javanese Home Gardens		●		●	●	●	●	●
New Guinea Agroforestry	●	●	●	●	●	●		●

of their particular regions. Yet from each there are useful lessons for the others: the intensive use of small plots, for example, or careful marketing can be applied to almost any tropical environment. Here we want to consider some of these key characteristics of sustainable farming.

Use of small plots. The intensive yet careful use of land is a hallmark of all promising agricultural approaches. For many this means producing the most out of a small piece of land. Javanese kitchen gardens, at less than one hectare per household, are exceptionally small, yet they can provide forty percent of a family's nutritional intake. In the market-oriented agroforestry of Peru, plots as small as one hectare can provide over eighty percent of a family's cash income. Even where land is less limited than in Java or Peru, small plots can be productive. Because all swidden-fallow systems depend upon regrowth of forest vegetation for their nutrient balance, a small plot in a forest may be preferable to larger ones or even to several contiguous farms. In addition, small plots and careful contouring, as seen in the New Guinea agroforestry system, can help reduce soil loss on steep slopes.

Successional schemes. Many successful farming systems, notably those of indigenous people, roughly imitate natural forest succession. The plot begins as a small opening in the forest, similar in size to a large tree fall; the farmer plants to imitate normal forest regrowth. For the first few years, root crops and annuals are tended before they are allowed to be overtaken by weeds. Perennial shrubs are tended and harvested for an additional few years, while longer-living trees are harvested for fruit, nuts, medicine, firewood and other household products. This system, and the agricultural yield, lasts in the fallow until it is cut again. As for pests and weeds, by the time invading weeds begin to choke understory crops, trees are developing and the closing canopy makes it difficult for them to grow. In many tribal systems where people have gained knowledge over hundreds of years, plantings are timed to coincide with seasonal changes in the forest such as the flowering or fruiting of native trees. Native vegetation, responsive to local weather patterns, is a much more sensitive indicator of growing seasons than adherence to a calendar. The four examples of neotropical indigenous agriculture, the Maya in Central America, the Kayapó in Brazil, the Bora and certain mestizos in Peru, show careful attention to local tropical forest ecology.

Diversity of crops. A common characteristic of sustainable agricultural systems is that they grow a variety of crops. Even the market-oriented schemes grow several products. The least diverse, the drip agriculture system in Mexico, still involves many types of vegetables, while indigenous farmers such as the Kayapó may cultivate or gather six hundred species of useful plants.

In many instances mixing crops enhances rather than inhibits growth. The use of nitrogen-fixing plants such as legumes in interplanting enhances soil fertility and produces more food. In Tomó Açu farmers, for example, use nitrogen-fixing trees instead of poles for supporting growing pepper vines. Beyond enhancing growth through intercropping, crop diversity holds other advantages. Where the invasion of pests and spread of disease can devastate a monoculture, farmers of diverse crops are buffered against total crop failure. With a wide range of crops it is also possible to harvest throughout a longer period of the year, so avoiding food shortages.

Gardens imitating forests. The branching structure, leaf arrangements and other features of tropical trees maximize their light-gathering ability. Those that require little sunlight, such as plants on the forest floor, receive as little as two percent of the light striking the outer edges of the canopy. Farmers can duplicate this structurally complex system in their gardens. Vegetables are planted in the herb and shrub layer; small understory trees produce fruit and nuts; larger canopy trees produce other fruit and nuts; vines and epiphytes yield other crops. Like the forest, the system is carefully balanced and highly productive.

Use of tree crops. Agroforestry, or the use of tree crops, is a relatively new term for an ancient practice. It is being touted as a superior form of agriculture, and for good reason. Tree crops tend to require less fertile soil, weeding and tending than other crops. Unattended, they will produce for many years. Almost all of the cropping systems presented in this chapter make use of trees, generally native species, for food, fodder, fuelwood and a variety of purposes including fiber, medicine, shade, support for other plants and soil conditioning through nitrogen fixation. Trees can be used for fodder to allow farmers to convert primary productivity to animal protein; for wood, especially from coppicing species which readily re-sprout; and for mulch, as in the intensive agriculture in Mexico, where leaf litter from neighboring forest is used on crops. This introduces more nutrients into the system, reduces water loss through evaporation, cuts back on weeding and, perhaps most importantly, encourages leaving small patches of intact forest around crop land.

Nutrient recycling. Nutrient cycling in tropical forests is rapid and efficient. Because most of the nutrients are tied up in living vegetation, when a plant or animal produces waste or dies, others quickly garner the nutrients released. Mulching and composting with plant parts not eaten or animal waste greatly enhances nutrient flow in a farm plot and may preclude the need for expensive commercial fertilizers.

Almost all sustainable agricultural methods include the recycling of plant and animal waste, and for good reason. With more nutrients, land can be cultivated more intensively and over a longer period of time without jeopardizing soil fertility. The result is more food per unit of land and less cutting of new land. Although more labor intensive, natural fertilizers are less expensive than commercial fertilizers. They are also more versatile. In New Guinea, leaves and inedible plant parts, by-products of processing coffee and rice hulls, are used as mulch. In the dairy cattle scheme, plant waste is composted with cattle manure and used to fertilize fodder plants. Waste from humans and domesticated animals are used by the Kayapó, Japanese farmers in Tomé Açu, and in Javanese kitchen gardens. In Mayan agriculture and Javanese kitchen gardens fish provide extra protein as well as fertilizer. The Kayapó use termite nests as a nitrogen supplement for soil, and the Mayans add nutrient-rich swamp soil to farm plots.

Attention to local conditions. On first glance it may appear that swidden fields are placed randomly, or for reasons of convenience. For the Kayapó, Bora and Maya, exactly the opposite is true. Careful placement of fields with attention to local topography for plantings is an integral part of these indigenous systems. Useful trees that could provide food, medicine or fiber are left standing, usually at the edge of the field, and crops are planted within the plot according to microclimate and soil conditions. Kayapó farmers condition the soil differently as they plant different crops. This requires intimate knowledge of soil types and possibly the identification of wild species as indicators, but this level of sophistication is not surprising: it is based on practices that have been perfected over hundreds of years. Even the simpler act of placing swidden fields next to economically useful trees, however, could contribute to a farm family's income.

Weed and pest control. The year-round benign conditions of the tropics favor populations of herbivorous insects, the spread of disease, and the rapid growth of non-useful plants. Intensive agriculture can provide an alternative to the widescale spraying of insecticides, fungicides and herbicides. Intensive systems take advantage of two resources available in most tropical areas: biological diversity and inexpensive labor. Intercropping, to take one instance, reduces the vulnerability of a particular crop to herbivorous insects by lowering its density. It also makes it more difficult for insects to locate the target crop; certain plants can actually help repel insects. Some provide a habitat for predators and parasitoids, which can help keep pests populations in check. Well-managed swidden-fallow systems also discourage pest and disease. The swidden constantly progresses through successional stages and hence changes as habitat for insects.

Weeds can be both controlled and used in traditional agricultural systems. In a labor-intensive system, the careful application of mulch and manual weeding can increase the length of time a clearing can be cultivated. And through experimentation, new disease and pest resistant plant varieties can be bred by allowing hybridization with wild varieties. The applicability of such systems to modern integrated pest management is obvious.

Inclusion of animals. Adding animals to a farm provides more protein as well as fertilizer. Fish-pond aquaculture is used by the Mayans and Javanese, and has recently been established in tribal lands in western Panama. Chickens and other domestic animals are fed scraps and their waste contributes to fertilizer. Wild animals also play a role. Attracted to the older fallows, they are hunted by the Bora, Kayapó and Maya. In Panama, the iguanas which feed on tree parts that are not used by humans are hunted and prized for their meat.

THE INTRODUCTION OF NEW SYSTEMS

To succeed, innovative farming must overcome two major challenges: it must produce more or better food than traditional farms and must be adopted by local farmers. These are difficult goals; subsistence farmers claim the poorest of soils, a fact which hampers increases in agricultural production, and farmers, like people everywhere, tend to resist new techniques. These techniques, therefore, must be fairly easy to learn, socially acceptable and inexpensive. A labor-intensive system may be adopted by people accustomed to hand weeding for example, but not by others used to low-effort farming. It is here that demonstration farms and agricultural extension projects, however time-consuming, can be most effective.

It would be pointless to promote the widescale adoption of any given farming method. Farming systems need to be adapted to local, highly varying, conditions, and the introduction of inappropriate crops and farming techniques can doom a new system to immediate failure. Inadequate testing or an attempt to transfer a successful technique from a dissimilar area can also be disastrous.

MARKET-ORIENTED AGRICULTURE

Market-oriented agriculture has its own set of requirements. The first is that the crops have a market. Secondly, the produce must be taken to the market or processing center, which requires adequate transportation. The crops must be able to withstand the trip and possible delays. Finally, any market-based economy is subject to the vagaries of market prices.

Marketing is the primary consideration for cash cropping systems, as the following case studies show. In the conversion of beef to dairy cattle, a marketing plan was developed simultaneously with the fodder-growing scheme. The farmers at Tomé Açu formed a cooperative to more effectively process and market their products. In other places, such as Quintana Roo, Mexico, cash crops have been easily introduced because markets and transportation already existed.

Lessons from Mayan Agriculture, Central America

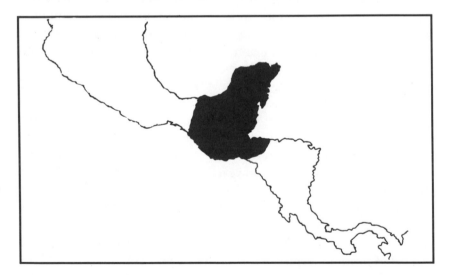

Among the best studied systems of lowland forest agriculture are those practiced by the Mayan Indians of Mexico and northern Central America. The Maya are of particular interest because they developed from a small population in forested areas into a major civilization with high population densities ranging from 100–200/km^2 in lowland areas of slash and burn agriculture to 700–1150/km^2

in regions where more intensive agriculture, involving raised beds, systems were developed.

Although there is considerable debate as to whether the high density and short rotations of the Maya's "milpas" (forest farm plots) contributed to the demise of their civilization through soil erosion and nutrient depletion, there is little doubt that they practiced ecologically sound farming and silviculture in lowland tropics. We can see what the Mayan system must have been at the height of Mayan civilization from archeological evidence and studies of current Mayan practices, particularly of groups such as the Lacandon of Southern Mexico, which use traditional methods.

The Mayan soil classification system was far more detailed and complete than that used by many local agronomists today. Plot sites were carefully selected based on soil type and certain indicator plant species. The standard "milpa" was a plot of forest or second growth that was cleared and burned, releasing nutrients into the soil from the ash. The Mayas then planted root crops and a mixture of other annual crops such as corn, beans, and squash. Now as then, these crops are mainstays, but in a typical Lacandon field, up to eighty crops are grown during an agricultural year. Milpas are not only diverse but productive: a milpa smaller than half a hectare may produce two and a quarter metric tons of corn and an equal amount of root and tree crops each year. Turned into pasture, this same land would produce only four kilograms of beef.

Milpas are highly labor-intensive and require continual weeding. The Lacandon commonly cultivate milpas for five to seven years until the crop production is not worth the effort to weed the increasingly aggressive second growth. As a testimony to the sustainability of this method of forest cultivation, it has been estimated that one farmer may clear as little as ten hectares in his lifetime. When a plot is finally abandoned, its soils are far more productive than those of most abandoned slash and burn plots, and it can be used for years after cultivation ceases. Its secondary vegetation will contain plants that are of direct use or attract animals, making the fallow fields almost like game-farms.

The planting and cultivation of trees is an important part of the traditional Mayan agricultural system. They planted leguminous trees to provide shade for cacao and coffee (a practice that has been generally discontinued) and milpas were cleared with tree management in mind. Frequently, preferred and sacred trees were left standing. Useful, fast-growing trees, characteristic of high light environments, were often trimmed back to a half-meter in height; after clearing and burning, they grew back quickly. Often a strip of forest was preserved adjacent to the milpa as a "seed" for the regeneration of the fallow.

In swampy areas of Mexico, Guatemala and Belize, pre-Columbian Indians planted various types of raised beds. The dominant agricultural method of the Aztecs in the Valley of Mexico, the chinampa, was created by digging narrow irrigation ditches on three or four sides of a small cultivated plot. Mud and soils from the trenches were continually added to the plot to create the raised bed. Swampy soils, often quite rich, helped maintain the nutrients in the plot. Beyond their use as irrigation ditches to control the water going to the chinampa, the trenches were used to raise fish and the banks are planted with useful trees. Recent research in the Mexican state of Tabasco has shown the efficiency of a modern chinampa system in maintaining high productivity of a diversity of crops, particularly when animals are integrated into the system.

Like other tropical agriculturalists, the Mayas also maintained kitchen gardens in their towns and settlements. These gardens included trees such as *Brosimum*, *Manilkara* and *Calocarpum*, for wood, fruits, seeds and forage, herbs, shrubs, vines and epiphytes (for example, chilies, roses, chayotes, orchids, maize, beans and onions) and many wild species. Forest trees were raised in the protective, elevated soil beds and then transplanted to form small forest stands of particularly valuable trees. Patches of old forest near Mayan ruins are often rich in useful plants and may owe their origins to the Mayan kitchen garden tree nurseries.

Mayan agriculture teaches us several lessons. First, it shows that it is possible to maintain a mosaic of forest and secondary vegetation that supports high-density human populations. Secondly, it demonstrates the usefulness of greenbelts and managed forest reserves which, when interspersed among agricultural clearings, help maintain diversity and the potential for forest regeneration. Finally, it indicates that even apparently pristine forest is probably the result of earlier intensive forest management practices.

SOURCE: Arturo Gomez-Pompa, University of California, Riverside, California, and James Nations, Center for Human Ecology, P.O. Box 5210, Austin, Texas 78763.

An Intensive Agricultural System for Forests with Karsted Limestone Areas, Mexico

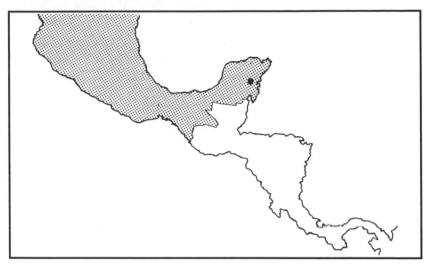

Much tropical forest has been destroyed to make room for agricultural and ranching operations of dubious economic value. In the Mexican state of Quintana Roo, which occupies the eastern side of the Yucatan peninsula, dry tropical forest is being cleared at a rapid rate to establish a highly subsidized cattle industry that has little real prospect of becoming economically viable. In addition, traditional shifting milpa agriculture, which is also widespread in this region, has led to extensive deforestation without raising the living standards of local subsistence farmers. In an effort to stabilize agricultural development in central Quintana Roo while assuring the survival of tropical forest, Felipe Sanchez and Patricia Zugasty have developed an innovative farming system that increases crop yields on very small plots. The ecological implications of this system compared to more traditional approaches to farming and ranching have been investigated by James Lynch and Dennis Whigham of the Smithsonian Institution.

Most of the Yucatan peninsula is underlain by limestone, which in many areas has weathered to form a pock-marked "karst" topography. A characteristic feature of this and other such areas is a lack of streams and other surface water. Rainfall quickly percolates through the fissured limestone bedrock to the water table, a process that is accelerated by the sparse soil cover that

typifies the Yucatan. In Quintana Roo, a family practicing traditional slash-and-burn milpa cultivation clears about two hectares
of forest each year. Given the locally desired fallow period of
about thirty years, each family will destroy some sixty hectares of
forest before returning to the original plot. The presence of a
three- to five-month dry season, exacerbated by the poor
moisture-retaining capacity of the thin, patchy soil, means that
only a single planting is possible each year. Despite the loss of
extensive areas of forest to shifting agriculture, crop production is
too low to produce even a modest economic return to the small
farmer.

To alleviate the dual problems of excessive deforestation and
poor economic return, Sanchez and his co-workers have introduced an intensive agricultural system that is ecologically sound,
affordable, well suited to the local climate and soil, and socially
acceptable to the Maya Indians who make up most of the farming
population of central Quintana Roo. Shallow wells are excavated
through the limestone cap rock, and small pumps are used to
bring the abundant groundwater to the surface. Crops are watered
continuously, using a simplified drip-irrigation system made from
inexpensive locally available materials. The irrigated plots, which
are less than one hectare in size and can be used for many years,
are mulched with leaf litter from the adjacent forest. Field trials in
two Maya villages revealed that the addition of forest leaf litter to
agricultural plots increased production ten to thirty percent over
that of control plots, depending on crop and soil type. The
positive effects of litter addition were strongest on soils that were
deficient in nutrients and humus-poor. Leaf litter is therefore
perceived to be a valuable resource, and farmers using the intensive system have a clear economic incentive to resist wanton
despoilation of the forest.

An intensive system has been actively promoted since 1984, and
has already produced impressive results. Not only have per-crop
yields increased several-fold over traditional methods, but the year-
round availability of water allows the planting of two or three crops a
year instead of the usual single crop. Many vegetables that are
difficult or impossible to produce by traditional farming methods
can now be grown using the intensive system. At the same time, the
long-term rate of forest destruction is reduced by a remarkable
eighty to ninety percent.

The government of the state of Quintana Roo has provided

funding to introduce the intensive system into twenty-three Maya farming communities, with the goal of encouraging the profitable production of vegetables for the state's growing urban market. This project was so successful that even after government subsidies were withdrawn most of the communities were able to maintain profitable production. The system has also been adopted by farmers in the buffer zone of the Sian Ka'an Biosphere Reserve in east-central Quintana Roo.

The advantages of the intensive system are even more striking when the system is applied to the production of forage for livestock. Cattle production in Quintana Roo is economically marginal, mainly because of poor pasture conditions (a stocking rate of one head per fifty hectares is the average). Nevertheless, cattle ranching has continued to expand rapidly, aided by government-subsidized loans. At present, ranching is the single most important land-use in the state. Large areas of forest are invariably felled and burned at the outset of a ranching operation, even if it is destined for bankruptcy within a few years.

In recognition of the pivotal role played by ranching, Sanchez and Zugasty have reoriented the original system toward the efficient production of livestock fodder on very small plots. Based on field trials, the system can produce enough fodder (corn, grasses, alfalfa) on a one-hectare plot to feed approximately sixty head of cattle on a continuing basis. Additional research is aimed at testing the feasibility of raising livestock other than cattle (for example, strains of sheep that are adapted to tropical conditions) with the aim of making small-scale meat production possible for individual landholders. Pilot trials involving the raising of sheep have been successful, and several large landholders in the Cancun area (northeastern Quintana Roo) have expressed strong interest in adopting the new method of fodder production as an alternative to massive forest clearing.

Ecological research has centered on two general issues: 1) the relative impact of the intensive-system (versus traditional agriculture and ranching) on birds, particularly overwintering North American migratory species; and 2) the possible long-term effects of litter removal on the growth and species composition of the tropical forest. A majority of local bird species, both migratory and resident, use the forest as a major habitat. The intensive system will greatly improve their long-term prospects by preserving large areas of forest that otherwise would be cut. The long-term effects of litter removal on the forest are more difficult to predict, but after three years of experimental litter harvesting there has been no detectable impact on tree growth or the rate of leaf production.

With the assistance of modest outside support for research and development, an ecologically rational and socially responsible agricultural system has been promoted and implemented. The key features of the system are that it encourages forest conservation, requires only modest capital investment, and can be easily adapted to the production of staple crops, vegetables and other cash crops, and livestock fodder. Most importantly to local farmers, the system has been shown to result in major increases in productivity and profits over traditional farming and ranching methods.

Sponsoring organizations: The system was originally developed with the support of the Centro de Investigaciones de Quintana Roo (CIQRO). Further research and development has been funded by the World Wildlife Fund.

SOURCE: Felipe Sanchez Roman, Puerto Morelos, Quintana Roo, Mexico, and James Lynch, Smithsonian Environmental Research Center, Edgewater, Maryland 21037.

Converting from Beef to Dairy Cattle, Costa Rica

Although cattle may not be the most ecologically appropriate animals for tropical America, it is unlikely that they will disappear from the agricultural scene. For environmentalists, the task is to curb the destructive effects of cattle ranching. One way of doing so is by making the conversion of forest into grazing land unnecessary. In Costa Rica, where cattle ranching is prevalent, the Tropical Agricultural Research and Training Center (CATIE) has done just that by introducing an ecologically rational, economically attractive alternative to beef cattle production in a small, planned agricultural settlement in the Atlantic lowlands of Costa Rica.

In 1980, CATIE began to promote the conversion of beef to dairy cattle in Cariari, a settlement that consists of about thirty 20-hectare parcels, which are moderate-sized farms by Costa Rican standards. Most Cariari families earned their livelihood primarily from beef production, grazing cattle on three-quarters of their

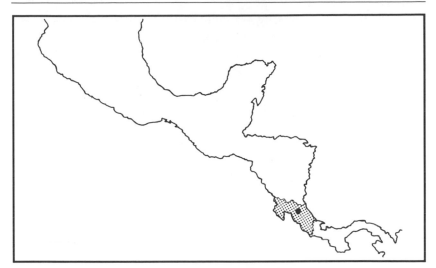

holdings. While beef cattle produce income, ranching has major drawbacks. The income is sporadic since cattle are brought to market only once or twice a year. Because about half of the beef is ultimately exported, its price is dependent on the world market. In recent years there has been a glut of beef and this has meant low prices for the Costa Rican ranchers. With prices constantly fluctuating, it is difficult to stabilize herd size at an economical, not to mention ecologically rational level. Furthermore, when the farmer finally goes to sell he often loses much of the profit to middlemen. This is particularly true for small farmers, who lack the acreage and the access to capital to raise cattle to full size and are forced to sell their herds to larger landholders who can fatten the animals for market.

Because of these drawbacks, and the ecological problems inherent in beef cattle ranching in the lowland tropics, members of the Animal Sciences Department of CATIE recommended that the farmers begin producing dairy instead of beef cattle. Selling milk products would provide a steady income, a stable market and a predictable price.

To succeed, the farmers needed rapid and easy access to market for their perishable product as well as an inexpensive alternative to commercial chemical fertilizers to make growing fodder economical. Access to the market in San Jose, Costa Rica's capital, was provided by a newly-paved connecting road. CATIE designed a simple composting system, using cow manure, to provide the inexpensive fertilizer for the fodder crops. This required that the composting system be built with readily available materials and with little labor, as labor is relatively expensive in Costa Rica.

The cow manure is placed in a shallow, plastic-lined pit and mixed with corn stalks and banana leaves. The corn stalks are harvested from the farmer's own field and the banana leaves are delivered free of charge from local plantations. The compost is ventilated through a system of bamboo tubes, obviating the need for labor-intensive turning of the material. As the compost stews, the heat it generates kills pathogens and weed seeds that would wreak havoc with crops. In only six weeks, the composted manure can be spread over the grass, feed crops and vegetable gardens. Controlled tests have demonstrated that this composted mulch produces fodder-crop yields that are comparable to those produced with chemical fertilizer, at a fraction of the cost.

Today milk is an important source of income for the people of Cariari. The development of this intensive crop and cattle system has reduced the need to clear new lands. The land held by the farmers of Cariari is being used to produce protein to meet domestic needs, rather than for export. Dairy cows convert fodder and grass of low nutritive value to nutritious dairy products. And, perhaps most importantly, herd size can be regulated and overgrazing reduced now that farmers do not depend on the fickle world beef market. In short, it is clear that tropical pastures can be managed soundly and that livestock, once a severe threat to forests, can be usefully integrated into sustainable agricultural systems.

Sponsoring organizations: CATIE (Tropical Agricultural Research and Training Center), Turrialba, Costa Rica.

SOURCE: M.E. Swisher, Pan-American Agricultural School, P.O. 93, Tecucigalpa, Honduras.

Iguana Ranching: A Model for Reforestation, Panama

In many tropical areas, rural people cannot meet their minimum protein requirements. This is caused, in part, by their dependence on slash-and-burn agriculture. This system usually leads to soil erosion, poor crop yields and general impoverishment. Moreover, wildlife, a traditional source of protein and other necessities for the rural poor, is often decimated.

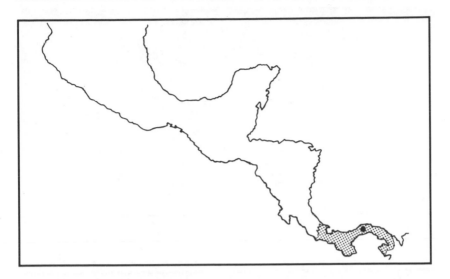

In Panama, one project is attempting to reverse these trends. The reestablishment of the green iguana (*Iguana iguana*) in iguana ranches will offer an important protein source and commodity for rural people of Panama and elsewhere. Compatible with forest preservation and reforestation, the project has developed a management scheme that promises economic viability using techniques which are culturally acceptable and easily learned.

Iguana is a traditional protein source throughout the neotropics. In many areas it has been virtually extinguished by overexploitation and habitat destruction. Most campesinos hunt iguanas and gather their eggs, but this is done without any thought of harvesting sustainably. A first step, therefore, in encouraging the sustainable use of iguanas is changing popular attitudes towards wildlife. Such a program was started in December 1985 by the Pro Iguana Verde Foundation in two communities. Campesinos are enthusiastic about iguana ranching and participate in experiments by rearing iguanas in their yards, collecting eggs from artificial nests, incubating eggs and censusing released iguanas.

Beyond their desirability as a food source for rural people in Latin America, iguanas have several biological features that make them attractive for management in wild, forested areas. They are large herbivorous reptiles that eat plant material otherwise unused by human beings or domestic animals. They have high ecological efficiency, which means that a large proportion of the food they eat is turned into meat, rather than used for maintenance costs such as

Iguanas, prized in Panama for their meat, are introduced into deforested areas along with useful trees in the iguana ranching scheme. The iguanas feed on leaves and people can harvest fruit and iguanas.

thermoregulation. In fact, an iguana eats roughly ten times less in a day than a bird or mammal of the same size.

On the other hand, iguanas' growth rate is slower than that of birds and mammals. Even with selection and enhanced nutrition, iguanas always grow slowly. Captive raising of iguanas, therefore, is much less efficient than raising chickens, for example. Compared to chickens, iguanas occupy the same space in a cage for a longer period of time, but achieve the same weight gain. Raising iguanas in cages until they reach harvestable size is probably not economically viable. Nonetheless, they are promising candidates for raising in unfenced forested areas.

Female iguanas are highly prolific, laying, on the average, thirty-five eggs a year and probably some three hundred eggs during their lives. In a stable, natural population only 2.5 percent of these eggs will hatch and survive to yearlings. With captive management, however, many more should survive, and it should be easy to build up stock for release into the forests.

Based on knowledge of the biology of iguanas, as well as the social and ecological situation in Panama, the project identified the following management objectives: to establish captive colonies of iguanas that would produce a predictable number of eggs; to select and breed iguanas with desired traits; to raise hatchling iguanas in captivity, protecting them from early mortality; and to release young iguanas into forests or woodlots in order to manage the lizards for protein production. Another aspect of the program was the development of an agroforestry scheme consistent with existing agricultural practices and with the needs of the iguanas. The trees that iguanas live in are also useful to humans for fuelwood, lumber and fruit.

The project thus far has met with a number of successes. The successful reintroduction and establishment of iguanas into depleted areas have been among the most important accomplishments. Hatchability of eggs and hatchling growth rates have been greatly improved through experiments determining the best incubation conditions and materials. The survival of hatchlings through their first year has increased from three to five percent to ninety-five percent. Good survival and growth rates were achieved at very high densities, up to thirty juveniles per square meter. This has been accomplished through experimenting with cage design, iguana density, and relatedness of hatchlings. Finally, by improving their nutrition, adult iguana growth rates have been greatly increased. It is possible that iguanas inherit certain features such as early reproduction and rapid growth; breeding these particular iguanas will boost future egg and meat production.

A landmark in this program was the first release of 1,200 captive-reared lizards on farms in the Panamanian countryside in December 1985. Ten months after the release, thirty to sixty percent of the iguanas could still be found around the release sites. Given the difficulties of relocating individuals, actual survival rates probably exceeded the sixty percent figure. Captive-bred iguanas are provided with feeding stations in the gallery forest, and experimental reforest-ation plots have been established to provide them with shelterbelts for soil protection, food and habitat. Now trees are being tested to see if they can survive when grown in conjunction with iguanas.

The management system is both easily transferred and economi-cally efficient. Due to increases in production and growth rates, iguanas now produce meat at about half the cost of feeding other domestic animals. Compared to cattle, for example, which also take about three years to raise, iguanas yield the same or even more protein per hectare. And, according to a best-case scenario from the pilot work, production of iguanas appears to be profitable even at the lowest sales prices. It is projected that a farmer who reforests his land and raises iguanas will be free of debts incurred by the project after eight to eighteen years. Moreover, iguana ranching does not require any degradation of land. Although it is still at a develop-mental stage, the iguana management project provides a realistic model for producing a source of protein in tropical areas consistent with sustainably managed woodlands.

Sponsoring organizations: W. Alton Jones Foundation and the James Smithson Society of the Smithsonian Institution supported the research. The Inter-American Foundation supports the educational and implementation aspects of the project. Cooperation was pro-vided by the Smithsonian Tropical Research Institute, the Natural Resources Institute and the Municipality of Panama.

SOURCE: Dagmar Werner, Smithsonian Tropical Research Insti-tute, P.O. Box 2072, Balboa, Panama.

Resource Management by the Kayapó, Brazil

The Kayapó Indians of the eastern Amazon Basin of Brazil employ a complex and sophisticated system of managing tropical forests. This system has been the focus of ethnobotanical studies by Darrell Posey and his collaborators since 1977. More recently it has become the object of interdisciplinary work examining all aspects of their know-ledge of the forests and savannas, including wildlife, plants and soils—knowledge which may provide models for ecologically sound management of diverse, lowland forest communities.

One important aspect of the Kayapó management system is that it begins with a complete and detailed "taxonomy" of ecological zones. In the savanna (or campo/cerrado) areas, the Kayapó recognize approximately fifteen formations that vary with the amount of shrub or tree cover, presence of water, and location with respect to hills and mountains. By contrast, modern ecological research in the tropics often relies upon cruder distinctions between "forest" and "second-growth." In addition, Kayapó women use a detailed soils classification system at least as sophisticated as that of most modern tropical agronomical systems in planning which crops to plant and where they should best be located. In all, the Kayapó recognize and use over six hundred species of plants.

Within the savanna there are small "islands" of forest. These too are managed by the Kayapó, who group them into seven types ranging from newly-formed vegetative clumps to forest corridors apparently maintained for defense. Forest islands are common close to Kayapó villages; as many as three-quarters of them may be human-made, and as many as eighty-five percent of the plants within them are thought to have been planted by the Kayapó. In all likelihood the current system is only a remnant of former trails and villages throughout the region, and of a time when human management of the campo actively determined the distribution of forest patches.

The Kayapó establish mulch piles in the forest islands. They then transfer them to small depressions in the campo, where they mix them with soil from termite mounds and portions of *Azteca* ant nests to form a mound one to two meters in diameter by fifty centimeters high. These mounds are nurtured over the years until they form islands as large as five hectares. Their functions are several: defense, shelter for the Kayapó when they remain in the campo, rest areas during the heat of the day, and the provision of privacy for other activities. Forest island plants are used for food (tubers, fruits, nuts, etc.), medicine, materials for baskets and other artifacts. Some food trees are cultivated to attract game. *Azteca* ants are introduced into the islands to repel leaf-cutting ants.

In the forest, the Kayapó practice slash-and-burn cultivation. The clearings are managed for a number of years, first for annual crops such as maize, beans, squash and manioc, then for perennial and tree crops such as sweet potato (which can be harvested for five years), yams (five to eight years), papaya (five years), banana (fifteen to twenty years) and Cupa (*Cissus gongylodes*) (up to forty years). The success of the system relies heavily upon continued nutrient enhancement of the soils through mulching and localized burning. Mulching and burning treatments are adapted for different crops. Old fields and secondary forest are particularly important for medicinal plants and as habitat for game. As in the Mayan, Bora and Taungya systems, the fallow land is managed by planting useful trees, some of which take many years to grow to harvestable size. The high concentration of fruit trees and relatively low vegetation attracts game, and clearings are frequently visited to gather seeds for transplanting into old fields and forest islands. The Kayapó understanding of crop succession is extraordinary. For example, the shade conditions of a fallow change dramatically as banana plants mature; the Kayapó know of approximately two dozen tubers and numerous medicinal plants that thrive under these conditions, known as "companions of banana."

The Kayapó used to travel over large areas along a network of trails, along which fruit-bearing trees were planted as a kind of way-station. One large Kayapó village, Gorotire, where much of this research has been done, still maintains a five hundred-kilometer trail system planted with yams, medicinal plants and fruiting trees. The villagers also plant in tree-fall gaps and openings created by removal of honey trees.

High-diversity gardens, rich in medicinal plants, are cultivated close to home. Older women maintain "hill gardens" of root crops, which, in the event of floods or other disasters, are important food sources. These gardens are planted in eight- to ten-year-old fallows.

Others are planted on outcroppings of basalt.

This research, along with work on Mayan agriculture and other indigenous agroforestry, calls into question the apparent "naturalness" of some existing tropical forests, and shows that a diversity of species and habitats can be maintained while yielding food, medicine and household products to their human managers. If the basic requirements of such a system—the systematic classification of soils and plants and the intensive management of second growth—are met, there is little to prevent the Kayapó's efficient and sustainable agriculture from being fruitfully applied in other parts of the world.

Sponsoring organizations: Conselho Nacional do Desenvolvimento Cientifico e Tecnologico, World Wildlife Fund International, National Geographic Society, National Science Foundation, the Ford Foundation and World Wildlife Fund US provided financial support for research. Other assistance was provided by Fundacao Nacional do Indio, Forca Aerea Brasileira, Museu Paraense Emilio Goeldi and the Universidade Federal do Maranhao.

SOURCE: Darrell Posey, Nucleo de Etnobiologia, Museu Paraense Emilio Goeldi, C.P. 399, 66.000 Belém, Pará, Brazil.

Japanese Farming in the Amazon Basin, Brazil

Japanese farmers have lived in the Brazilian Amazon for over fifty years. In their small enclaves, they practice a diverse and sustainable form of agriculture. Until recently little was known of their culture or indeed of their successful farming practices. Now researchers are looking at the town of Tomé Açu, one of the largest Japanese communities in the region, located 210 kilometers south of Belem and the mouth of the Amazon, to see why Japanese farmers have flourished where most other colonists have failed.

The Tomé Açu farmers face the same ecological constraints plaguing all farmers in the upland forests of the Amazon: high rainfall, soil containing few nutrients, and a high concentration of metals in the earth. Their farms are located in much the same places as other farmers' plots. What, then, makes them unique? Researchers say that it is their attitude towards the land, and hence their cultivation practices that distinguish these farmers from the others.

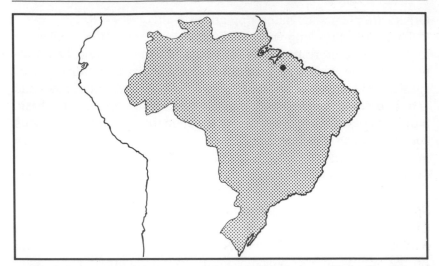

In Japan, a small amount of land supports a high density of people. The villagers at Tomé Açu, apparently following this tradition, view land as a limited resource that requires careful tending. Their agroforestry, for example, mirrors the pattern of natural succession. After a plot of land is cleared and burned, it is carefully cultivated. Fast-growing crops such as rice, cotton and beans are interplanted with intermediate-lived perennial vines such as black pepper (*Piper nigrum*) and passion fruit (*Passiflora edulis*), followed by interspersed plantings of tree crops such as cacao and rubber. Tree crops, in turn, are used as substrates for epiphytic crops (those that use other plants for support but not nutriment) such as vanilla bean orchids. One farm product is used to nourish another. In the case of animals, rice straw from chicken coops is spread around the base of fruit trees as a mulch and supplementary fertilizer.

As in many indigenous systems, the farming at Tomé Açu is time-consuming and physically taxing. But here the expenditure of labor is cut by environmentally sound agricultural techniques that may increase yields over the long term. Plants are pruned and culled to control pests and weeds, and are left in the plot to decompose, or are burned and returned to the plot as ash. Other plant materials, such as the residues of black pepper processing, are spread for insect control. Many of the problems of crop management are solved by careful interplanting that results from continual experimentation. Interestingly, no single system of crop planting and rotation is used by the roughly two hundred farmers that live in Tomé Açu.

The community is well organized to process and market its cash crops. Although the land is held in moderate-sized (twenty to sixty hectares) single family plots, certain crops such as vanilla and cacao are processed on the farm for added value before being marketed outside the farm. Products such as latex (for rubber), passion fruit and oil-palm nuts are processed and marketed cooperatively. The farmers employ extension agents and agricultural economists to help determine which crops to plant and to track changes in commodity prices. Their processing and marketing strategies add considerable value to their cash crops and may generate the cash needed to pursue their capital-intensive system.

The farming systems at Tomé Açu and those of other Japanese settlements in the Amazon are still poorly understood. However, their success demonstrates that through continued innovation, combining traditional and modern techniques, agroforestry systems can be highly productive on relatively small amounts of land, even with poor Amazonian soils. Given the cultural legacy of the farmers at Tomé Açu and the amount of capital required to start farming, it may be difficult to transfer their system to other settlements. But there is clearly much to learn from this approach to farming in the Amazon.

Sponsoring organizations: National Science Foundation and the Jessie Smith Noyes Foundation; cooperation from Museu Paraense Emilio Goeldi, Belem, Brazil.

SOURCE: Scott Subler and Christopher Uhl, 202 Buckhout Laboratory, Pennsylvania State University, University Park, Pennsylvania 16802.

Long-term Cultivation of Swidden-Fallows by Bora Indians, Peru

Traditional agriculture in the tropics often involves the felling and burning of a swatch of forest, known as a swidden, followed by the planting and tending of one or two crops before the plot is left to return to forest. Practitioners of such systems are constantly on the move, clearing new patches of forest or re-clearing fallows that have regenerated over many years. The entire process is usually regarded as land-expensive, sustainable only in sparsely populated areas where fallow land could remain undisturbed. In fact, this form of

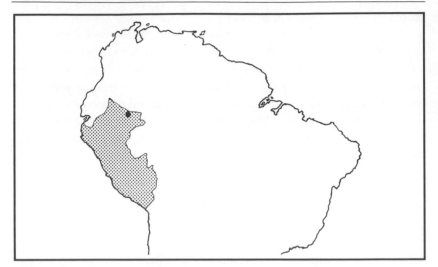

indigenous agriculture may be a model for the development of modern sustainable systems. It is far more productive than we have been led to believe, and, as the swidden system of the Bora Indians of the Amazon shows, it is both sustainable and productive.

The Bora village of Brillo Nuevo, on the Ampiyacu River near Iquitos, Peru, is surrounded by a mosaic of fields and forests. Although the land immediately around the village has been cleared at least once during the fifty years the Bora have lived at Brillo Nuevo, within a few miles of the village the forest is essentially untouched. Only the area around the village is worked. Although fields cleared from old forest are generally more productive, the regenerated, newer forest is closer to the village and the river. Thus the land around the village moves repeatedly through a cycle of swidden and fallow of about ten to twenty years, although some forty-year-old forest fallows can be found.

Careful management is practiced from the very beginning of the cycle. As in Mayan agriculture, the Bora are quite discriminating in how they cut the swidden. They leave valuable timber trees and palms near the edges of the clearing, and place manioc cuttings throughout the field along with plantings of pineapple, fruit trees and other crops. The crops are planted in the clearing in patches, partly in response to topography and the distance to the edge of the clearing, but also in response to soil types. Coca is planted near trails, peanuts are planted where manioc has been harvested; and certain crops are carefully nurtured with ash from wood fires. Not only do the fields have a distinct structure, but they differ considerably. There appear to be two basic cultivation strategies: low-diversity fields that grow mainly manioc and pineapples and higher diversity fields, which are more likely to receive the careful long-term management.

The third year is pivotal for the management of a fallow field. Manioc is usually not grown after two years, but fruit and other tree crops are beginning to produce. The field is still weeded periodically and small patches of pineapples and other crops are cultivated. Over the next few years, fruit crops will be harvested and the coca carefully weeded, but the secondary vegetation gradually takes over the plot. During the next ten years, the Bora concentrate on clearing around the fruit trees as part of the swidden gradually becomes an orchard and produces fruits throughout the year. The secondary vegetation is a source of valuable medicinal substances and construction materials, ranging from palm fiber to the bark of certain vines. Animals are attracted by the fruit, and the fallow becomes a good hunting-ground. After thirty years, important trees can be felled and floated down river to the market.

The Bora fallow management system is oriented toward subsistence agriculture; with the exception of some timber and craft items, little is sold at the distant market. Undoubtedly some cash crops could be added to the system. With or without cash crops, however, the essential soundness of the Bora's cultivation techniques, and those of other traditional societies in Amazonia, indicates that deforestation is not an inevitable result of swidden-fallow agriculture.

Sponsoring organizations: UNESCO Man and Biosphere Program, the National University of the Peruvian Amazon in Iquitos, and the University of Wisconsin, Madison.

SOURCE: William Denevan, Department of Geography, University of Wisconsin, Madison, Wisconsin 53706.

Market-Oriented Agroforestry in the Amazon, Peru

Most of the well-studied agroforestry systems practiced in the lowland neotropics are employed by tribal groups for their own subsistence. This leads to the general perception of traditional sustainable agroforestry systems as primitive, romantically appealing, but

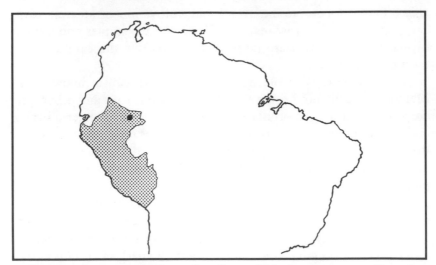

perhaps a less practical mode of tropical development. Running counter to this notion are recent studies of non-tribal "mestizo" farmers living near Iquitos, Peru. These show that the Peruvian farmers engage in cyclical and sustainable agroforestry that is quite commercially successful.

This agroforestry system is found only thirty kilometers southeast of Iquitos and is therefore within easy access of the large markets in that city. Agroforestry is practiced by the some two thousand inhabitants of Tamshyiyacu. Located along the main channel of the Amazon, the village is supported by some cultivation of seasonally flooded, or *varzea* lands as well as the upland, or *terra firme*, forest. In addition, fields and forests provide food and materials for handicrafts and charcoal which are sold in the market. But the activity that produces the greatest bulk of the cash income is the cultivation of swidden-fallows. Based on data obtained through interviews with Tamshyiyaquino families it has been estimated that approximately two-thirds of their annual cash income comes from cultivated fruit such as umari (*Poraqueiba sericea*), cashew and inga (*Inga edulis*), and another one-fifth from annual or perennial crops such as manioc, rice, plantain and papaya. Other commercially important products include game, charcoal, palm fiber hammocks and baskets, and a small amount of forest fruits and medicinal plants.

As elsewhere, the cycle for a "typical" field begins with the clearing of a patch of forest. At this point the system differs from other traditional systems: rather than burning the slash indiscriminately, larger woody vegetation is converted into charcoal which is sold at market. The first crops consist of annual and semi-perennial crops which are partly replaced in the second year with the planting of tree crops. During the initial five years, the intensively managed crops are phased out and the trees, principally umari and Brazil nuts, take over. Products from trees can be harvested for the next twenty-five to fifty years. Maintenance consists of protecting them from invading livestock and weeding. Weeding, required frequently in the first few years, is gradually reduced to an annual or semi-annual event. When fruit and nut production finally declines, the larger trees are harvested and converted into charcoal. The field is allowed to lie fallow for about six years before the cycle begins again.

These fields are managed to produce a great quantity of commercially important crops and materials, as well as supplying products for household use. This gives a family a substantial and dependable source of income. The diversity is enhanced by the management of several fields at different stages. Initial data suggest that the Tamshyiyaqino farmer's annual income was, in the early 1980s, as high as $5,000, compared to the local average of about $1,200. The villagers practicing this system make a better living than many of those growing subsidized monocultures such as upland rice and pineapple. Although many details of the system are still unknown, these studies indicate that an agroforestry system that retains many of the features of indigenous systems can provide immediate economic benefits to practitioners, while acting as a model of ecologically sound and sustainable agricultural development.

Sponsoring organizations: Research was conducted through the UNESCO Man and Biosphere Program with a grant from the USDA Forest Service.

SOURCE: Christine Padoch, New York Botanical Garden, Bronx, New York; J. Chota Inuma, Universidad Nacional de la Amazonia Peruana, Iquitos, Peru; W. De Jong, Agricultural University, Wageningen, The Netherlands; and J. Unruh, University of Wisconsin, Madison, Wisconsin.

Javanese Home Gardens, Indonesia

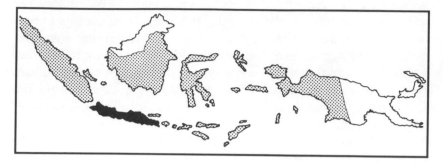

People in the tropics have farmed successfully under crowded conditions for millenia. Although the possibility of continuous agricultural development is based partially on ecological conditions, traditional cultivation systems allow for the intensive use of a limited amount of land. These systems are being studied as potential models for the sustainable use of cultivated lands in other areas.

One such intensive form of land use is the kitchen garden, found around the world. It is known by various names, such as the Malayan kebun, the Chagga garden of Mt Kilamanjaro, the jardin creole of the West Indies, the dooryard garden of Central America, and the pekarangan of Indonesia but they all have much in common with modern agroforestry systems: they are multistoried, and contain a diversity of cultivated species which makes them reminiscent of tropical forests. Since they achieve sustained yields without significant inputs of fertilizers and pesticides, these gardens represent an attractive alternative to shifting agriculture.

Home gardens are agricultural systems in which multipurpose trees and shrubs are managed in conjunction with annual and perennial crops, and animals, within the yards of individual houses. The entire crop–tree–animal system is intensively tended by family labor. Subsistence is the principal function of most home gardens, but surpluses may be bartered or sold for cash.

In Java, Indonesia, home gardens are known as pekarangan. They are one of the most important sources of food on this ancient rice-growing island. Prehistoric settlements in Java may date six to ten thousand years ago. Agriculture was likely developed quite early, probably beginning with shifting cultivation; rice has been cultivated for approximately two thousand years. Although comprising only seven percent of Indonesia's land area, Java contains sixty-two percent of its population. In 1980, an average of 690 people lived in each square kilometer. Since most rural Javanese are farmers, this means that the average farm size is low—about 0.6 hectares per household. The result is severe poverty and malnutrition.

The Javanese believe that they are part of their environment and that any damage to the environment will adversely affect them. Natural resources are exploited carefully. Overexploitation is avoided since the search for excessive material wealth is considered sinful. Such beliefs have led to ecologically sound agricultural practices including contour plowing and the terracing of hillside fields. However, with a rapidly growing island population, greater pressure on uncultivated lands, and evolving cultural attitudes, ecologically sound practices have not always held sway. Erosion and deforestation have occurred.

The typical Javanese village is virtually hidden by the trees planted in home gardens. Varying in size from a hundred square meters to two or three hectares, home gardens contain anything from ground creepers to tall trees. On the ground layer, vegetables such as spinach, beans, cucumbers, tomatoes are found along with various medicinal plants; their richness and density depend on the amount of light reaching the ground. In the second layer, there are foods such as taro, cassava, banana, papaya and salak, and ornamentals such as frangipani. The canopy also has distinct levels. The lower layer may consist of small trees such as citrus, guava, coffee or cacao, while the intermediate layer may contain fruit trees such as jack fruit, mango, rambutan, leguminous trees, sugar palm and bamboos. The emergent layer of the canopy, which sometimes reaches thirty-five meters, may contain durian or coconut.

Erosion is well controlled in the home garden. Plant litter is left in the garden, protecting the soil and cycling nutrients, and the multistory structure helps dissipate the effects of strong rains. This is often in marked contrast to the surrounding countryside, where agricultural lands may be heavily eroded. The differences are particularly striking in the dry season when lands outside the village are dessicated and brown, while village gardens remain green.

Recycling is efficient. Plant litter is saved, except in front of the house, where there is a play area for children and a place for adults to gather. In West Java, home gardens commonly contain fish ponds. The fish are fed with table scraps, kitchen leftovers, cassava and taro leaves. Chicken coops, horse stables, and toilets are often built above the pond and animal and human feces and urine flow directly and indirectly into it and feed the fish. When the fish are harvested, the pond is drained and the mud is removed and mixed with goat or sheep manure, composted and used to fertilize plants. The more traditional the village, the less the material exported from the system and thus the greater the recycling. With a growing reliance on a market economy, however, it will be increasingly difficult to recycle. Also, with greater population densities, pollution of drinking water sources by fish ponds has become increasingly common, with a concomitant spread of disease.

Home gardens contribute significantly to the production of the Javanese household. Gardens can provide twenty to twenty-five percent of net household income, while requiring only about seven percent of total labor. While gardens do not generate as much gross income as rice production, net incomes are higher because of the high cost of rice production. The gardens also contribute substantially to villagers' subsistence providing as much as forty percent of their caloric requirements. Unlike rice, the production of the home garden is year-round and provides crucial cash and nutrition on a regular basis.

SOURCE: Otto Soemarwoto, Institute of Ecology, Padjadjaran University, Bandung, Indonesia.

An Extension Service for Shifting Agriculturalists, New Guinea

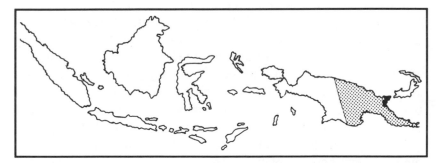

In many tropical countries, shifting agriculture is the single greatest cause of deforestation. In Papua New Guinea, for example, roughly eighty percent of its three million people live as subsistence farmers. The government has actively encouraged the growing ranks of urban poor to return to their villages and to clear more forest land for crops. Because forest plots become unproductive quickly, farmers must move their fields every few years, and the cycle of clearing, burning and deforestation continues.

As long as forest remains, it acts as an economic safety valve. Subsistence farmers can carve out an existence in forests without further taxing the thinly spread social services of the cities. Yet small-scale agriculture is limited. Large areas of Papua New Guinea are tied up in timber concessions, and its best agricultural land supports coffee, oil palm, rubber plantations and other export crops. This, coupled with intense slash-and-burn practices which are creating wastelands of Kunai (*Imperata cylindrica*) grassland over the foothills of the country, suggest that shifting agriculture will be rapidly eliminated as an option for the small farmer.

Traditional systems of slash-and-burn agriculture involve careful long-term management of the land. These agroforestry systems are not often used by the country's contemporary practitioners of shifting agriculture, in part for social and economic reasons, but also because they do not have access to the detailed knowledge of traditional farmers. This is increasingly true as more and more farmers are recent immigrants from remote villages and urban areas. In order to provide information about sound agricultural practices, ecologists from the Wau Ecology Institute in Morobe Province of Papua New Guinea developed the Shifting Agriculture Improvement Program.

In 1976, scientists at Wau began experimenting with simple techniques for making forest farm plots last longer. The focus was on two limiting aspects of farming in tropical areas: soil erosion and pest attacks. Several techniques, relying on easily procured material, were developed. They included the heavy use of compost and mulch, bedding techniques such as contour mounding to reduce erosion, intercropping of fruit and timber trees, and integrated pest management. Since the project was begun in a coffee-growing area, a scheme for maintaining high soil fertility relied on compost of coffee-hull mulch, mixed with weeds and grass. Composted mulch was formed into long mounds following the contours of hillsides to reduce erosion. Each compost mound supported at least three different crops, including one species of legume to maintain high levels of soil nitrogen. Existing trees were either left standing or others planted to stabilize the soil and recycle nutrients. An alternating planting scheme was employed where successive crops were planted in compost mounds in old furrows, and old mounds became the new furrows.

The agroforestry extension service in Papua New Guinea emphasizes simple techniques such as mulching with compost to increase soil fertility. Based on an illustration from *Subsistence Agriculture Improvement Manual*.

By the late 1970s this system was being demonstrated to farmers in Morobe Province. It is still employed today. Certain of its features have been spread throughout the country by Lutheran, Catholic, Baptist and other mission workers involved in rural development projects. The approach was broadened and summarized in a simple book on sustainable techniques, known as the Subsistence Agriculture Improvement Manual, which became the basis of an education program for rural assistance workers. Since village women are largely responsible for cultivation and care of forest farms, the extension program relies on women teachers.

Sponsoring organizations: Provincial Government of Morobe Province and Wau Ecology Institute.

SOURCE: Wayne Gagne, Departments of Education and Entomology, Bishop Museum, Honolulu, AI 96817.

Chapter Three
NATURAL FOREST MANAGEMENT

In terms of disturbance to the land, natural forest management lies between strict reserves and agricultural clearing. Natural forest management appears to be a compromise: it acknowledges the need for an economic return from tropical forest, while still preserving the environment. It is widely practiced in temperate areas. In the United States natural forests cover 77 million hectares, only seventeen percent of which is classified as strict wilderness. Certain species of forest organisms suffer under managed conditions, but a vast majority can be protected. In the tropics, the impact of management is likely to be more severe. The diversity and complexity of tropical forest ecosystems makes them more vulnerable. However, managed forest is less harmful ecologically than unsustainable agriculture and cattle ranching. It is certainly far superior to degraded land.

For the most part the purpose of forest exploitation has been quick financial returns, with consequences ranging from the damaging to the devastating. By contrast, sustainable forest management shows smaller short-term gains: the techniques are costly; the systems often require leaving some land untouched. But economic as well as ecological returns are possible through careful forest management. In the long term, the benefits are likely to equal or exceed those from farming and ranching. In this chapter we present seven projects that represent three basic financial benefits from natural tropical forests: timber, non-wood products and water and energy from watersheds. None of these approaches has achieved the goal of developing a large-scale, ecologically sound management system.

Wood products usually are the first to come to mind when considering economic uses of natural forests and many good systems exist for managing temperate forest ecosystems. The problems with management for timber products are two-fold. First, logging operations themselves are often highly damaging, and this is becoming more of a problem with increasing mechanization. Heavy machinery compacts soil and in most operations, the vegetation surrounding the trees being harvested, is also damaged. Up to two-thirds of the non-target trees in some areas are damaged or destroyed when marketable trees are extracted. Ultimately this can destroy young individuals of economic tree species and preclude the regeneration of the forest. Logging roads and skidding trails further contribute to soil compaction and erosion. The second major problem lies in the decline in economic value of forested land after the most valuable trees have been removed. There is little incentive to manage land of marginal economic value and forested land that has had the best timber cut often is converted to other uses.

SUSTAINABLE WOOD EXTRACTION

The sustainable practice of timber harvesting is far less common than it should be. Techniques exist to exploit tropical timber without degrading the land, but they are rarely used. In fact, it appears that the great dipterocarp forests of the Old World tropics, primarily found in Malaysia and Indonesia, can be managed fairly easily. Some of the most sophisticated management techniques were developed and practiced in Southeast Asia for many years, but today few commercial operations practice sustainable forestry. Although we do not provide case studies of specific sustainable forestry projects from Southeast Asia, it is worthwhile outlining the principles of that region's promising approach.

The most widely accepted example of tropical silvicultural success was developed in the dipterocarp forests of peninsular Malaysia. Foresters began practicing "regeneration improvement fellings" to increase the amount of seedling regeneration in these forests. Based on observing the effects of more widescale and less controlled logging of these woods during World War II, forest managers discovered that the seedlings found on the forest floor responded well to the extensive removal of canopy trees. This could be due to the reproductive pattern of dipterocarp trees, called masting, in which trees of many species flower and set seed in the same year. This produces large, even-aged stands of trees. So rapid was the regrowth in artifical clearings that the Malaysian Uniform System (MUS) was developed. It was determined that over a sixty-year period, given the right treatment, a forest could be logged, regenerated, abandoned (apart from clearing out of vines) and then harvested. The system consisted of cutting large economically useful trees and poison-girdling other trees greater than five to fifteen centimeters in diameter. The result was a relatively uniform growth of seedlings in the genus *Shorea* that could all be logged and marketed. It is this structural and taxonomic uniformity that gave rise to the name of this silvicultural treatment. The MUS also involved a series of timber surveys that assessed those tracts of land that had sufficient regeneration for new harvests. Many tracts did not have sufficient seedling densities, so recent modifications to the MUS have included active seeding to make regeneration more predictable.

The Malaysian Uniform System was practiced widely in the lowland forests of peninsular Malaysia from the early 1950s to the 1970s. Although this management scheme seemed to work well, other economic demands on the forest land, particularly for farms and plantations, all but completely removed these forests, obviating the need to manage them. Therefore, although the MUS appeared to be a sustainable silvicultural scheme, it was not in practice long enough for the second rotation cuts to be made. The focus of silvicultural work has shifted to the hill forests, particularly in Borneo.

Most of the silvicultural work in the Malaysian states of Sarawak and Sabah and in Indonesian Kalimantan is recent and its effectiveness has yet to be evaluated. Several factors prevented widescale adoption of the Malaysian Uniform System for hill forests. For one, there is a less predictable carpet of seedlings available for the next cut in these forests. In addition, the steep topography makes the exact abundance and composition quite variable, and so silvicultural treatments must also be more variable. The topography makes logging difficult, and there is a greater chance of damage to regenerating seedlings. Although the MUS is still practiced in a modified form, the new emphasis is on selective management systems.

Selective silvicultural systems rely on the management of young trees rather than of seedlings. More trees are left standing, and treatment consists of poisoning undesirable species and cutting vines. One common form of silvicultural treatment in these selective systems is known as "improvement thinning." The aim is to poison-girdle and cut trees and lianas that directly compete with economically valuable species. The resulting gaps in the forest are kept small to reduce the invasion of pioneer trees and vines that choke regrowth. Test results from these plots have been mixed.

The sustainability of logging operations can be increased by relatively simple measures including the reduction of extraction damage. And although the effectiveness of various silvicultural treatments are not at all clear, there are theoretical reasons to believe that some of these treatments could be cost effective. As a practical matter, the survey work necessary to determine the nature of regeneration and future marketable trees requires a great increase in the number of people with technical forestry skills. The paucity of trained foresters has resulted in the reduction of silvicultural surveys and the separation of logging and silvicultural treatment. In Sarawak and Sabah there is a growing backlog of managed lands that have not been treated after logging.

Natural forest management can be profitable if the costs of treatment are kept relatively low. Economic benefits can be realized, but only in the long term: the per-acre value of timber extracted in this fashion is low compared to other potential economic uses. At the same time, however, it is possible to retrieve non-wood products from managed forests. It would be natural to develop extractive industries in conjunction with low-level natural forest timber extraction.

Conservation purists may balk at advocating natural forest management as a way of conserving tropical forest. And it is true: silvicultural treatment inevitably changes the composition and structure of forests. Some of the systems, such as the MUS, create forests with greatly simplified structures. Selective logging systems in the hill forests of Malaysia, for example, tend to increase the proportion of non-dipterocarp trees. We would argue that, as a complement to natural ecosystem preservation, forest management offers an opportunity to promote forest cover while supplying at least some revenue.

Higher species diversity and associated difficulties in marketing many types of wood make New World tropical forests appear to be less conducive to sustainable wood production. As the projects described in this chapter show, however, the problem is surmountable. These projects, all in South America, rely upon careful extraction techniques and attempt to simulate natural gap formation and regeneration. Cutting cycles range from thirty to forty years, and gaps are formed near intact forest, which then provides the seed for regeneration. At Palcazu the cut areas are roughly the size of large natural gaps, although Bajo Calima has had successful regeneration in extremely large openings. All three operations avoid the use of heavy machinery and use draft animals or overhead cables to move the sawn logs out to collection points. At Bajo Calima even the roads for logging trucks are constructed to minimize soil compaction and erosion. Parts of trees that are not needed, such as leaves, bark and branches, are left to decompose on site, returning nutrients to the forest ecosystem. The Celos System includes selective management to increase the density of economic tree species by poison-girdling potential competitors, and these are also left to decompose. All three systems were developed as the result of detailed investigations into natural forests, and ongoing research assesses the progress of regeneration.

Table 5. Qualities of Natural Forest Management

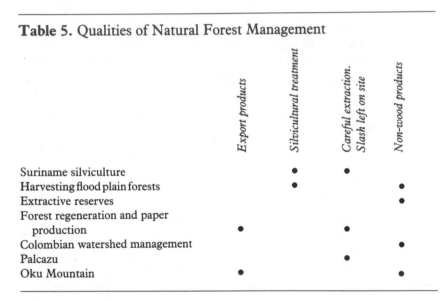

	Export products	Silvicultural treatment	Careful extraction. Slash left on site	Non-wood products
Suriname silviculture		●	●	
Harvesting flood plain forests		●		●
Extractive reserves				●
Forest regeneration and paper production	●		●	
Colombian watershed management				●
Palcazu			●	
Oku Mountain	●			●

NON-WOOD PRODUCTS

Minor forest products, or non-wood products, are collected by tropical forest peoples around the world. Non-wood products supplement agricultural produce, form the basis for cottage industries and probably account for millions of dollars of foreign trade each year. There is very little documentation of the financial value of tropical forest non-wood products, but this is changing as more governments and financial institutions discover the potential of minor forest products. The careful extractive use of tropical forest may provide economic returns and support wildlife. However, unwise collection practices can be quite destructive. Like natural forest management for wood, exploitation must be controlled and allow the resources to recover between harvesting. Future study is required to determine how to expand the non-wood uses of tropical forests and include them in regional development plans while avoiding overexploitation. As in the case of Mt Oku, the value of careful extraction methods is often realized too late.

Non-wood products include rattan, bamboo, fibers, latex, gums, resins, oils, medicinal plants, nuts, fruit, and animals and their products. Some of the commercially important species such as rubber are grown in plantations, but most are still collected from natural forest. By increasing the value of natural forests to agricultural communities, these products may in turn encourage forest preservation.

The non-wood products that result from the three harvesting schemes discussed in this chapter are sold on local markets, but there is no reason why they cannot be exported. The key to successful harvesting of non-wood products is to avoid overexploitation. Rubber tappers, for example, do this by rotating the use of three trails of rubber trees and by making careful cuts. The Caboclos have developed a detailed knowledge of forest ecology and diversified their use of the forest. Certain species, encouraged in forests close to markets, are discouraged in less accessible areas. By harvesting a diversity of species they are less likely to overexploit any one species. Similarly, the people of Mt Oku declared a moratorium on *Pygaeum* bark collection when it was overexploited. Now they are replanting and planning better harvesting techniques. They are also developing cooperatives to make and market honey, handicrafts and other products.

ENERGY PRODUCTION

The Rio Nima watershed protection project is included in this chapter rather than with the other reserves because of its emphasis on energy production. The reserve was established to protect the watershed and stem erosion that caused problems with a hydroelectric dam. The fact that the project is being administered by an electric company reveals that energy can be profitably extracted from natural forest, and that forest preservation can be cost-effective.

A Sustainable Silvicultural System for Forests, Suriname

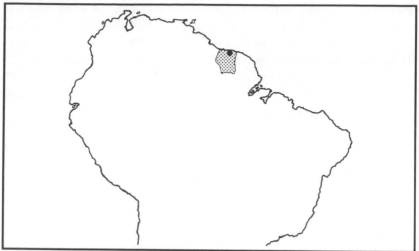

Much of the forest in the temperate zone is not protected, but is managed and selectively logged for timber. Although the degree to which such logging disturbs habitats is a constant source of debate, management of natural forests presents itself as a land use option lying between strict protection and permanent clearing for other economic uses. While selective logging has been a common activity in tropical forests, particularly in Asia, the long-term sustainability of this practice is a point of contention.

Two major problems arise with repeated exploitation of tropical forests. First, commercially marketable trees are often scarce and as they are harvested the value of the forest declines. Logging quickly becomes economically unattractive. Secondly, the forests suffer ecological degradation from logging machinery and from indiscriminant opening of the canopy, which creates opportunities for "weed trees" and vines to overgrow the plot. Both of these problems can be overcome through proper harvesting and manipulation of the plot, but such manipulations must be tailored to the particular forest type that is being worked. The development of these schemes requires long-term ecological studies.

Researchers from the Agricultural University at Wageningen were able to conduct long-term studies to develop a harvesting and silvicultural system for the upland forests of Suriname. Unfortunately, political unrest has terminated most of the research and the system has not yet been applied to the Foresty Zone of one and a half million hectares, designated for timber extraction by the Suriname government.

Suriname, located on the north coast of South America, has a relatively small population, concentrated near the coast. The government has already promoted some selective logging by criss-crossing

Damage to tropical forests from timber extraction is minimized in the sustainable silviculture scheme in Suriname, by careful planning of cutting trails and the use of winches for felling.

the forest with an extensive road system. Selective logging has generally been *ad hoc* and the quality of the forest for further timber extraction has been continually degraded. The forest has also been depleted of nutrients. As in many upland tropical forests in South America, the soils are highly acidic, and leaching has further stripped the soil of nutrients. With most of the nutrients tied up in the vegetation itself, large-scale removal of trees could result in an acute loss of nutrients.

The most serious damage to the forests generally comes from skidding and other logging procedures. Through careful planning, cutting removal trails and increasing the use of winches, the area damaged by logging can be reduced from twenty-five to twelve percent. Any increased costs due to this careful extraction is more than compensated by the greater efficiency of running the machines on an organized skid-trail system.

It is a considerably greater challenge to design an economically viable harvesting program. As in many other lowland tropical forests, a major impediment is the scarcity of commercially valuable trees. In Suriname approximately fifty species are exploited, and they generally only comprise about ten to twenty percent of the forest. Without any management, the growth rate of commercial species is slow. To some extent, this can be offset by encouraging the lumber industry to use more types of species. The major thrust of the work in Suriname for the past twenty years has been the development of a harvesting and management plan that reduces environmental damage and increases the number of commercially valuable trees in a stand.

The Celos Silvicultural System (CSS) is the result of long-term research on 120 hectares of experimentally managed forest. The project continues its research on special plots, where it assesses the effects of the harvest and carefully maps and censuses both commercially desirable and undesirable trees. Harvesting on these plots is "polycyclic," which means that a series of small fellings are conducted at twenty- to twenty-five-year intervals, rather than one large harvest. The modification of the stand comes from three interventions after felling during the twenty-year rotation, during which large undesirable trees are poison-girdled and large vines cut. Although commercially undesirable trees are killed, they are left to decompose in the forest ecosystem—the only nutrient loss in the ecosystem is that from the harvested trees, and even here, the gaps created by the harvest allow regeneration. A final important feature of the CSS is that even though its silvicultural treatments are uniform, the resulting forests are still heterogeneous in structure and composition.

If carefully followed, the Celos System will make selective logging more efficient and productive, and therefore more economically viable. It will also produce a forest of mixed age and species composition, reduce the threat of fire, conserve nutrients, maintain habitat for forest wildlife and allow extraction of other forest products while only minimally changing the forest's watershed protection abilities.

Sponsoring organizations: Anton de Kon University of Wageningen; the Wageningen Agricultural University; Suriname Forest Service and the UNESCO Man and Biosphere Program.

SOURCE: N.R. de Graaf, Department of Silviculture and Forest Ecology, Agricultural University, P.O. Box 342, 6700 AH Wageningen, Netherlands.

Harvesting the Floodplain Forests, Brazil

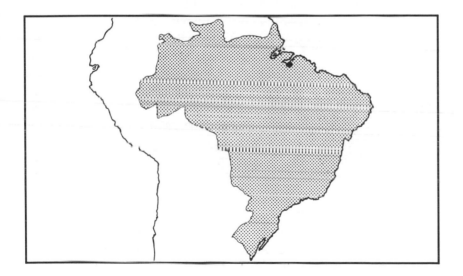

Harvesting available forest products is probably the least ecologically damaging economic use of tropical forest lands. Although forest-based economies are commonplace among indigenous groups, it is less usually found among the descendents of European settlers. The Caboclos, the rural mestizo people of Amazonian floodplain forests, provide a striking exception, as they have been managing and using forest resources of the Amazon estuary for well over a century. Since 1984, the Museu Goeldi in Belém has sponsored a multi-disciplinary team to uncover the secrets of this management system.

The floodplain forest, covers 25,000 square kilometers in the Amazon estuary, and is a unique ecosystem, quite distinct from the upland forests that cover most of the basin. Its limited distribution within the Amazon, coupled with its close proximity to the river systems and relatively good soils, make it a prime spot for agriculture. These forests, with their low tree species diversity and a high concentration of economic tree species, are also ideal for extractive management.

The Caboclos actively manage their forest resources. They thin or eliminate undesirable species to favor economically useful species, making the complex decision of what to favor or exclude on the basis of various economic trade-offs. For example, at one rural community studied on the island of Saracá, local residents purposely eliminated the açai palm (*Euterpe oleracea*). Ironically, this species is valuable as an important source of palm hearts and as the base for a staple beverage. When it grows close to markets, the açai is intensively managed and harvested, but transportation becomes impractical from more remote sites. At these less accessible sites, the Caboclos manage the forest understory for cacao. There, the açai palm is selectively eliminated because its dense root system threatens growth of the cacao saplings. The Caboclos plant their cacao seedlings around the base of another palm, the buriti (*Mauritia flexuosa*) to encourage growth. Buriti collects detritus at its base, forming a nutrient-rich bed for the young cacao trees.

The inhabitants of the floodplain forests do not rely solely on managed forest, but maintain plots in the context of a broader land-use strategy. Each village has areas of managed forest which they develop through selective regrowth of swidden fields and house gardens or from selective thinning of previously unmanaged forest. In addition to conventional crops, these practices yield a high variety of forest-extracted products such as edible fruits, fibers, timber, palm hearts, game, organic fertilizer, honey and medicinal plants.

The Caboclos' resource strategies and techniques are being studied at three sites in the Amazon estuary. Detailed studies of managed and unmanaged forests have also been undertaken. The

Caboclos' system is also a subject of a comparative research program at the Museu Goeldi, which includes two other Amazonian groups, the Kayapó and Kaa-por Indians.

Sponsoring organizations: Museu Goeldi, the Ford Foundation, World Wildlife Fund-US, and the Brazilian National Research Council (Conselho Nacional de Desenvolvimento Cientifico e Tecnologico-CNPq).

SOURCE: Anthony Anderson, Division of Economic Botany, Museu Goeldi, Caixa Postal 399, 66.000 Belém, PA, Brazil.

Extractive Reserves: A Sustainable Development Alternative for Amazonia, Brazil

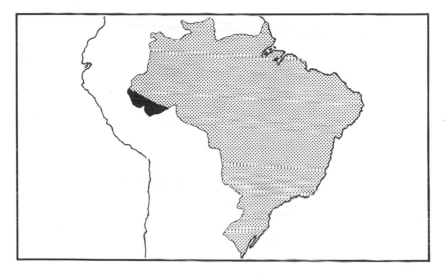

The proposal to establish reserves for the harvesting of extractable forest resources combines forest preservation with a workable development effort. In Brazil, it is a plan with a strong local constituency as well. Proposed in 1985 at the first annual meeting of rubber tappers in Brazilia, it is supported by an emerging community of autonomous rubber tappers and other users of extracted resources, such as brazil nut gatherers. This is a group of potentially great political importance as there are a half of a million rubber tappers and an unknown number of other "extractivists" in the Amazon

Basin. The proposal is gathering support from various Brazilian government ministries and international development agencies and the first reserves may be declared as early as 1988.

The proposed reserves address the need for forest preservation in rapidly developing parts of the Amazon, particularly in the Brazilian state of Acre where the rubber tappers have their strongest and most visible organizations. Extractive reserves offer a mode of forest use that is both immediately economically competitive and sustainable in the long-run. In Acre, for example, although the estimated current per-hectare value of extractive production is somewhat less than that for cattle ranching and agriculture, its value is increasing more rapidly than agricultural and beef. In 1980 the value of rubber, Brazil nuts and several minor products in the State of Acre was over $26 million. More importantly, reserves used for extraction will continue to produce over the long run. This is an opportune time to implement such a proposal because of the impending paving of the Rondonia–Acre road.

The proposal for extractive reserves would give land-use rights to "free rubber tappers," those tappers who work as independents. Today these autonomous tappers have only limited land rights and are often faced with expulsion in the face of ranching or colonization projects. In the current climate of land speculation, land rights are a potentially explosive issue. In 1960, 97 percent of the land in Acre was devoted to extraction, with 88 percent of this land concentrated in seven percent of the holdings. Most of the tappers were essentially indentured workers. By 1970, because of the rise of ranching and other agriculture, only 65 percent of the land was held for extraction, while the number of rubber holdings had increased ten-fold. Most of this increase was due to the growth of autonomous tappers, those who began to use land without paying rent or having any legal title. Each free rubber tapper family works around three hundred hectares of land. Tappers ideally work three trails, each containing around 150 to 200 rubber trees. Only one trail is visited per day so that the individual trees are tapped only once every three days. To avoid harming the trees, one cut is made per visit and the size of the patch where cuts are made and distance between patches are carefully measured.

The organization of cooperatives and the use of small processing plants has allowed tappers to sell their product directly to industry rather than to a middleman processor. The position of free tappers is precarious: they are caught between their increasing debt to middlemen and the constant threat to their livelihood from forest clearing for ranching and logging. The growth of political power is an important step toward the achievement of land tenure and the preservation of the rubber tappers' way of life.

Ideally, a rubber tapper cuts each tree once every three days to collect latex without harming the tree. Extractive reserves will allow people in the Brazilian Amazon to earn income without disturbing the forest ecosystem.

Cooperatives have been organized with the assistance of two organizations: Acre Pro-Indian Commission (CPI), which works with indigenous groups, and Projecto Seringueiro (Seringueiro is the Portugese word for rubber tapper), which organizes in regional rubber tapper communities. Recently, the Brazilian government recognized land claims by indigenous groups in Acre. These groups now control a significant amount of land in Acre and are committed to a mixture of economic activity, including agriculture, hunting and fishing, as well as extraction. CPI cooperatives have been an important organizing force and have helped some Indians escape debt peonage by establishing community marketing. The project has also been providing social services, although most of the cooperatives still require outside financial support.

Similar work is being conducted with neobrazilian rubber tappers in Projecto Seringueiro, which emphasizes education. Its cooperatives began with support from Oxfam, UK, and now have government sponsorship.

The rubber tappers were organized by a national agricultural workers' union (CONTAG), and have been demonstrating against forest clearing since the mid-1970s. The strategy has now turned towards halting deforestation altogether and convincing the Brazilian forest service (IBDF) to enforce conservation laws that are on the books.

Perhaps the most important step toward establishing extractive reserves has been taken by the newly-formed National Council of Rubber Tappers. Its proposal for reserves is being discussed by the Brazilian government and international development agencies, and the Institute for Agrarian Reform and Colonization has agreed to halt its practice of dividing and expropriating areas of forest that are under extractive management. Two areas of Acre are slated for extractive reserves and others are being planned for Rondonia.

Sponsoring organizations: Support for cooperatives is provided by Oxfam, UK. Research has been supported by World Wildlife Fund-US and Instituto de Estudos Amazonicos.

SOURCE: Mary Helena Allegretti, Instituto de Estudos Amazonicos, R. Itupava 1220, 80040 Curitiba, Panara, Brazil and Stephan Schwartzman, Environmental Defense Fund, 1616 P Street NW, Washington, D.C. 20036.

Natural Forest Regeneration and Paper Production, Colombia

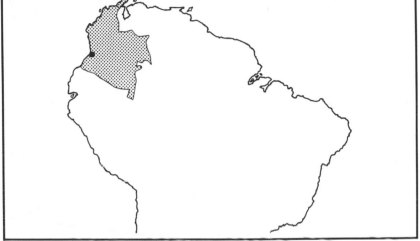

For nearly thirty years, the corporation Carton de Colombia SA has been demonstrating the feasibility of renewable harvesting of wet tropical forests. The project, located in the Pacific lowlands of Colombia, involves clear-cutting areas of forest in a manner that causes minimal damage to the soils and the conversion of trunks and large branches into pulp for paper products. The logging causes little damage to the soil because trees are cut individually and the logs are removed with aerial cables. In addition, this commercial concession has supported extensive research into the regeneration of natural forest vegetation.

The actual procedure is simple but effective. Trees are felled with axes or chain saws, then cut up into 1.5 meter lengths and debarked with a machete or axe. Larger branches are cut for their timber; smaller branches and foliage are left to decompose. The wood to be processed for pulp is piled up along paths, then lifted and removed with skyline cables, stacked along roads, and trucked 140 kilometers to the processing plant in Cali. Longer logs are hauled off to lumber yards by independent truckers. To permit truck access in these very wet clay soils, roads are constructed from rock ballast laid on non-woven geotextile placed on the soil.

Although sustained forestry began in 1960, the 60,200-hectare Bajo Calima concession was officially ceded to Carton de Colombia by the National Institute of Natural Resources and Environment

(INDERENA) in 1974. Half the concession area is currently under management. In the management area, approximately one-fifth of the forest is protected from timber extraction because of steep slopes; another one-tenth is placed in strict reserves without road access. The managed site is divided into work areas which are further subdivided into sixty plots of 600 hectares each, of which two are harvested each year on a sustainable-yield management cycle.

The company has consistently monitored the pace and nature of regrowth since 1974. Regeneration is rapid, diverse and lush, partly because the topsoil is undisturbed and seeds and fruits are abundant in the soil immediately after harvest. Ninety-eight percent of the regeneration is from seed, so it is important that seed dispersal by mammals and birds continues unimpeded. To this end, patches have only been clear-cut when they are less than a kilometer from standing forest.

The concession's duration is thirty years, a figure arrived at by calculating annual tree diameter growth at one centimeter per year. Researchers at the concession are studying the forest cover and species composition of thirty-year-old forest. They have found a similarity of tree species composition and about half of the tree volume of the primary forest by the fifteenth year. Although pioneer species, including the genera *Miconia*, *Vismia* and *Cercropia*, are the most common in four-year-old regeneration, by the fifteenth year tree species composition was more similar to that of the primary forest.

The most detailed studies were conducted on two-year-old natural regeneration after clear-felling. Although there were major changes in relative abundance of species, composition was more similar, seventy-seven tree species were found in both the primary forest and the regenerating plots. Forty of the primary forest species were not found in the two-year-old plots. It is likely that this is partially a function of rarity of some of the trees and partially a problem of sampling, because many of these forty trees were found elsewhere on regenerating plots.

As in most management schemes, the project's success depends upon restricting the use of the land to one activity. In this sense, the Bajo Calima project is fortunate in being located in an area of extremely high rainfall (7,500 mm) and poor soils, which makes the land undesirable for most agriculture. The logging concession also provides employment, although this can be a drawback. The roads provide access to forest, trees are illicitly removed for timber, and the sustained forestry plan is undermined. The Bajo Calima concession demonstrates that the extraction of timber from patches or forest can be conducted sustainably. However, the sustainable practice is severely threatened by illegal timber removal from the concession.

Sponsoring organization: Carton de Colombia SA.

SOURCE: William Ladrach, Zobel Forestry Associates, P.O. Box 37398, Raleigh, North Carolina 27627.

Public and Private Cooperation in Protecting and Managing a Tropical Watershed, Colombia

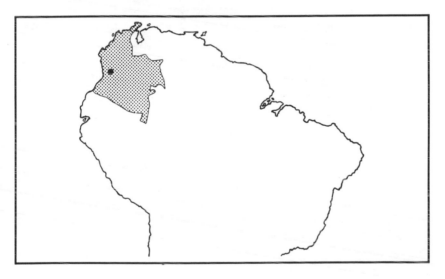

Colombia depends heavily on its torrential Andean rivers for drinking water, irrigation and hydroelectricity. Deforestation and erosion now threaten these water resources. Careless environmental management and poorly-controlled agricultural expansion have dealt a serious blow to crucial watersheds and to the country's diverse wildlife, which includes more than 1,700 species of birds. Undoing the damage is a task which has required both private- and public-sector efforts.

The Rio Nima Valley is an area of critical importance for the protection of the water and energy resources of Palmira, a city of 200,000 at the base of the Andes. For the past forty years, people have settled on the slopes of the Rio Nima watershed and converted them to pasture and cropland. The result has been a loss of forest and considerable erosion. The Rio Nima watershed project was created in response to this urgent situation. Sponsored by the city of Palmira, the Cauca Valley Corporation (CVC) and the Fundacion Natura, it aims to reforest cut-over slopes and to protect the remaining natural forest. Much of the funding for the project comes from consumers of the water and hydroelectric power which the Rio Nima provides.

A unique feature of this project is that it involves cooperation between local and national government agencies, conservationists and, most unusually, a power utility. In Colombia, a number of semi-autonomous corporations manage watersheds for hydroelectric projects, much as the Tennessee Valley Authority does in the USA. The largest of these corporations, the CVC, has produced hydroelectric power in Colombia's Cauca Valley for thirty years. It also runs a system of parks and forest reserves in the Cauca Valley which are among the best managed in Colombia.

About 1,200 people live in the Rio Nima watershed. The growth of agribusiness in the valley has pushed many of them into the surrounding hills, where, in order to survive, they cleared land for cultivation, pasture and dairy cattle. As a result, the CVC estimates, a quarter of the land is moderately to severely eroded.

The first-step towards watershed conservation was an agreement between the municipality of Palmira and the CVC. They took a reconnaissance survey of the valley, and then, in 1973, the CVC took charge of protecting the forest. Nonetheless, land degradation continued and in 1980, Palmira, the CVC and Fundacion Natura developed a comprehensive watershed management plan to reverse this trend. With support from The Nature Conservancy and Fundacion Natura, the committee developed a cooperative plan for management of the upper watershed.

The plan covers both the reasons for, and the consequences of, erosion. It seeks to provide land titles and technical assistance to people living in the lower valley so that they will not be forced to move to the hills. Where land has already been degraded, replanting is in progress to replace vegetation cover and provide industrial wood for Palmira. The project also provides credit for planting non-native commercial trees such as cypress, pine and eucalyptus, and an extension service teaches farmers simple conservation methods.

Only about a quarter of the high watershed hills are still covered by natural forest; a tenth is forest plantations, a fifth is natural pasture and a third is pasture created by forest clearing. Dairy farming is the major economic activity; other types of agriculture use only about eight percent of the land. Ultimately, the project hopes to have a watershed with a quarter of its land in natural forest cover, just over half in managed or plantation forest and less than a quarter in cultivation.

Since the CVC started protecting the natural forest, incursions have dropped by ninety-five percent. In 1985, it was determined that the highest slopes (2,100 to 4,000 meters) of the watershed were not appropriate for any type of economic activity, so land holdings in this area are gradually being purchased. As of 1987, a thousand hectares of reserve lands had been added to Paramo de las Hermosas National Park.

Sponsoring organizations: Empresas de Palmira, Cauca Valley Corporation, Fundacion Natura and The Nature Conservancy.

SOURCE: John Shores, The Nature Conservancy, Latin American Program, 1785 Massachusetts Avenue NW, Washington, D.C. 20036.

Sustained-Yield Management of Natural Forests in the Palcazu Development Project, Peru

It is a well-known fact that a small amount of continual disruption is actually necessary for tropical forests' regeneration. The Palcazu project in the eastern Andes of Peru, designed by the Tropical Science Center and supported by USAID, is an attempt to replicate some of the features of natural forest disturbance as a basis for a sustained-yield forestry system. Because of its proximity to the Amazon Basin, the project has also had to tackle the problems typical of that region, particularly colonization and the protection of native peoples' culture and land rights.

Approximately six thousand people live in the Palcazu Valley, of which slightly over half are Amuesha Indians. The rest are colonists and resident ranchers of German, Swiss and Austrian descent. The Amuesha live in twelve native communities, where they cultivate

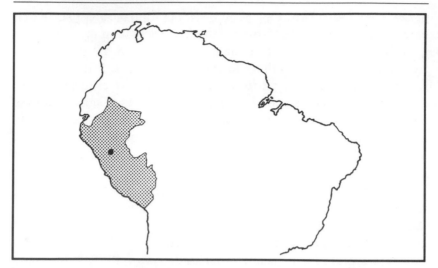

manioc, maize and rice using a swidden-fallow rotation. Helping the Amuesha to get land titles has been a key accomplishment of the wider development project of which the Palcazu project is a part, the Pichis-Palcazu Special Project. (Unfortunately, this project has promoted road construction—a potential first step towards colonization—which somewhat offsets the environmental benefits offered by the Indians' new land rights.)

Although the valley retains approximately three-quarters of its original forests, there has been significant clearing along the rivers. Most of the valley is unsuited for agriculture; rainfall is extremely high (up to 7,000 millimeters a year) and the soil, mainly acidic red clay very low in nutrients, is notoriously poor. On the basis of these soil and slope characteristics, the project divided the lower valley into three categories: land that can support some agriculture and grazing (about thirty-five percent), land that should remain in protected forest (eighteen percent) and land that should be placed under active forest management (the remainder).

Many attempts at natural forest management in the lowland tropics have failed. Yet recent trends may make such management an attractive use of tropical forest lands. Perhaps the most important development has been economic. National markets, which traditionally valued only a small number of the finest tropical woods, are now open to profitable trade in generic tropical woods, often marketed under such rubrics as "common oak."

The Palcazu management system involves harvesting small strips of forest, twenty to fifty meters wide and of variable length, in thirty- to forty-year rotations. Each harvested strip must be at least two hundred meters from those cleared in previous years, and for each block of five strips, an additional section is set aside to act as a reserve for old growth vegetation. Four or five sections of strips are under management at any given time. Those cut in 1985 have already produced a rich regrowth; like natural tree-fall gaps, the surrounding vegetation ensures that the cut strips will have plenty of seeds for regrowth.

During the harvest, major branches and trunks are removed, while small branches and foliage are left behind to provide nutrients. Draft animals do the work of taking the wood out of the forest; this is not only inexpensive, but far less damaging to the soils than machinery. The remaining vegetation is not burned, so that nutrient recycling occurs slowly through decomposition.

The timber is processed by a small, cooperative-owned sawmill in the valley. The Yanesha Forestry Cooperative, an organization of five Amuesha villages, has included a thousand hectares of forest in its management plan and may ultimately expand to manage 40,000 hectares. Each Amuesha community hopes to begin at least one rotation by 1989. The cooperative is already a model for other development projects; Amuesha leaders and Tony Stocks, the project anthropologist, have been asked for advice by similar projects elsewhere in Peru.

The harvested wood is sawn into logs, posts (which are treated with preservatives to extend their life and increase their value) or converted to charcoal. In 1986, one small saw mill could process about twelve hectares of forest a year, and a larger processing center consisting of a saw mill and pole treatment plant (PresCaps) was completed in 1987. The AID-funded road into the Palcazu Valley passes the wood-processing plant and extends to the Iscozacín River. Current estimates by the Tropical Science Center suggest that potential net profits after wood processing can be $3,500 per hectare worked, with a thirty- to forty-year rotation, although the actual harvest results have yet to be seen. If profitable, the Palcazu project will be one of the first examples of a small-scale sustainable forest management scheme in lowland tropical forests that relies upon indigenous manual labor.

A potential problem for this and other local development projects is that the main road into the valley may not be maintained well enough to be a reliable conduit for commercial goods. As a general concern, it may prove difficult in the current economic climate to provide sustained funding to firmly establish the project, and it will need to survive a difficult period when the AID-sponsored experts leave.

Sponsoring organizations: USAID and the Tropical Science Center.

SOURCE: Gary Hartshorn, Roberto Simeone and Joseph Tosi, Jr, Tropical Science Center, Apartado 8-3870, San Jose, Costa Rica.

The Conservation of Oku Mountain Forests for Wildlife, Watershed, Medicinal Plants and Honey, Cameroon

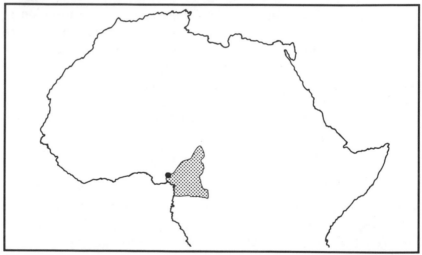

Without the montane tropical forests on the slopes of Cameroon's Mount Oku, local villagers would have no livelihood. They harvest the medicinal bark of the *Pygaeum* tree to sell to drug companies, and earn cash by keeping bees that produce honey from the nectar of forest flowers. Other people, too, are dependent upon the forest. In the lowlands, agriculture is based on the water that flows from the watershed protected by the forest. These peoples' livings, and their sustainable use of the forest, are being threatened by the encroachment of slash-and-burn agriculturalists, who bring with them goats and sheep that overgraze cleared land, invade the forests and increase the chance of wildfires. Although the area has been cleared over the past twenty years, recent road improvements in the area have exacerbated the situation; the destruction of the remaining 7,000 hectares of forest is imminent.

The Mt Oku forest is particularly important because the Cameroon highlands are populated by animals unique to the area, including two species of birds. In the past few years, a two-pronged effort has been devised to reduce the threats to the remaining forest. First, efforts have been made to classify the forest as a protected area, so preventing major disturbance while allowing sustainable collection of forest products. Secondly, local industries that rely upon forest-based products are being encouraged and improved.

The honey industry is growing, with the membership of a local honey/handicraft cooperative almost tripling in the past few years. The sale of honey provides cash for a third of local villagers, although their processing and marketing needs improving. Reliable marketing is also a problem with the stools, drums, masks and other wood carvings that the village men sell through the cooperative.

The major extractive industry associated with the Oku forest is the harvesting of the bark of *Pygaeum africanum*. Used to produce a drug that helps control urinary problems associated with enlarged prostate glands, the bark is processed and marketed commercially by international pharmaceutical companies. Bark collectors are issued permits by the Forestry Department, which controls where they can operate and limits how much bark they can harvest. Theoretically, *Pygaeum* bark can be harvested sustainably in the following fashion: the trunk is divided into four longitudinal sections and two opposite quadrants are stripped of bark. The trees with their two remaining sections of bark intact are left to rest for four years before the next harvest. Under the old practices, however, trees were overexploited and killed, and the harvesting of bark is now halted until new trees are established.

The *Pygaeum* replacement project began by engaging the support of the people of the village of Oku. The proposed project was discussed extensively with the traditional village council. One approach to the problem is to promote local forest-related industries. Local villagers are also interested in the prospect of developing agroforestry on the degraded lands, using native species. Seven thousand *Pygaeum* seedlings, grown in nurseries, are ready for planting at forest borders and on burned and cleared lands. This replanting, combined with stricter regulation of harvest techniques, could allow the lifting of the moratorium on bark extraction and the development of a viable extractive industry. Other plans are being developed to strengthen the Oku honey cooperative. An extension service on apiculture, forestry, agriculture and grazing lands is also planned.

Sponsoring organizations: International Council for Bird Preservation, Missouri Botanical Garden, World Wildlife Fund, USAID and the Ministry of Agriculture, Cameroon.

SOURCE: Michael Rands, International Council for Bird Preservation, 219c Huntingdon Road, Cambridge CB3 0DL, UK, and Duncan Thomas, Missouri Botanical Garden, P.O. Box 299, St Louis, Missouri 63166-0299.

Chapter Four

TROPICAL FOREST RESTORATION

Restoration is not a substitute for the preservation of tropical forest. It is time-consuming and expensive and can never fully replace original vegetation. In tropical forests, the incredible diversity and complexity of the ecosystem make restoration particularly difficult. But despite this, and although it addresses a symptom of deforestation rather than the disease itself, restoration ecology is worth serious consideration. It can speed regeneration in managed systems, make non-productive land productive again and protect closed-canopy tropical forest. As the projects and research discussed in this chapter show, restoration is at its best when applied to the regeneration of cut forest or to the reclaiming of barren land, rather than to the task of planting tropical forest from seed.

The restoration of degraded land is different from regenerating natural forest. The primary goal of most restoration projects is to take degraded land and make it productive. It is a strategy that is adopted in areas of severe erosion and soil compaction, where quick action is desperately needed. Normally fast-growing exotics, rather than native trees, are planted. Regeneration of natural vegetation, by contrast, is a protracted process. And while restoration projects generally get funding for implementation, regrowth of natural vegetation is still largely a focus of research.

AIDING REGENERATION OF NATURAL FORESTS

Increasingly the landscape of tropical forests consists of vast areas of cleared or shrub-covered lands dotted with relict patches of forest trees. In many areas, preserving the forest has become an irrelevancy; too little exists to be preserved. However, these small patches of forest can be rejuvenated so long as the original mixture of plants and trees exists. In fact, this may be the only option in areas of almost total decimation. Because we usually have little understanding of the original species composition, this approach is not wholly satisfactory, but it is better than nothing.

Research therefore plays an important role in the development of regeneration as a method. Its greatest successes are based on a thorough knowledge of the factors that limit establishment of plants in the anthropogenic clearings, coupled with an inventory of the composition of the original forest. With such an understanding, few interventions may be necessary to re-establish natural vegetation.

Table 6. Qualities of Restoration Ecology Projects

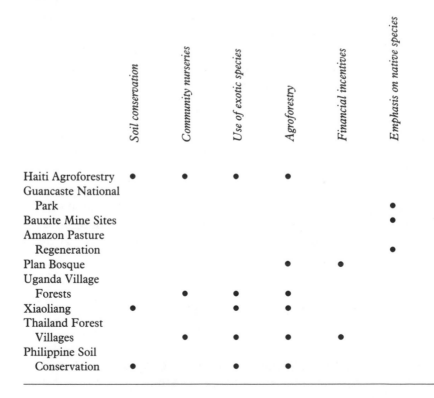

	Soil conservation	Community nurseries	Use of exotic species	Agroforestry	Financial incentives	Emphasis on native species
Haiti Agroforestry	•	•	•	•		
Guancaste National Park						•
Bauxite Mine Sites						•
Amazon Pasture Regeneration						•
Plan Bosque				•	•	
Uganda Village Forests		•	•	•		
Xiaoliang	•		•	•		
Thailand Forest Villages		•	•	•	•	
Philippine Soil Conservation	•		•	•		

There are examples of how such an innovative effort can be piggy-backed on an existing research program—on, for instance, Janzen's work in Guanacaste National Park or Knowles' work in regenerating forest trees on mining scars in Brazil. Both these examples will be fully described in the case studies which follow; the point here is that, as they will show, regeneration can be imaginatively and successfully applied to even the most troublesome environments.

Despite its high cost, Knowles' project is considered a reasonable expense in the context of the operating budget of a mining company. But it is sobering to realize that mining companies have high operating budgets, and the costs of reclamation may be prohibitive for other commercial ventures such as cattle ranching. On a per-hectare basis, this type of reafforestation is quite time-consuming and expensive, and would probably not be feasible for extensive tracts of land.

GROWING TREES ON DAMAGED LAND

Regeneration is especially appropriate in a dual environmental strategy of improving the economic value of already degraded lands while preserving existing intact forests. A number of the projects we discuss later are aimed at developing some sort of economically useful tree cover on cleared and abandoned lands. Since reforestation is given high priority by development agencies, we have also included case studies of current tree-planting and soil conservation projects. Only occasionally is it an explicit goal of such projects that the restoration of degraded land reduces incursions for wood into existing forests. We looked for projects that had a tropical forest conservation component in their restoration programs and found three: in Uganda, Ecuador and Thailand. But there is no reason why the two goals cannot be combined. By improving timber production on abandoned lands, and thereby increasing community appreciation of natural forest, tree planting programs could serve to save tropical forest.

Most restoration programs depend heavily upon a few species of non-native trees because they are convenient and require less knowledge of the original flora. These include fast-growing eucalypts, conifers (pines and cypress), nitrogen-fixing trees and hardwoods such as teak, all of which can be sold commercially for timber or pulp after a relatively short growing period. For conservationists, these plantations of fast-growing exotics are generally undesirable.

They tend not to support indigenous wildlife, and like other monocultures may be more susceptible to pests and diseases. Subsistence
farmers, moreover, who most need trees on their land, cannot wait
ten to fifteen years for the trees to grow to income-producing size.
Industrial production of such wood may not benefit people quickly
enough to reduce pressure on natural forest. It is becoming increasingly common, therefore, to grow trees that produce some products
immediately, as well as future timber. Here, restoration projects
must compromise between two conflicting approaches. A simple
agroforestry system employing a small number of common species is
easy to develop and transfer. On the other hand, projects tend to be
more successful when they are based on innovative combinations of
native species. Perhaps the best approach, as in the case of the
Haitian and Ugandan projects, is to start with the easiest systems
and then gradually diversify the tree species, a process which, by
giving farmers almost immediate results, ensures long-term community support for the project.

Often the goal is more basic: to establish soil on eroded hillsides.
In these areas, the priorities are stabilizing the soil and establishing
fast-growing, nitrogen-fixing trees. Crops can be planted in soils that
collect in hedgerows and contour fences made of fast-growing trees.
The trees that act as soil-stabilizers can, in turn, be used for animal
fodder. This system provides both immediate and lasting benefits for
practicing soil conservationists.

Several projects in this book underscore the need for continued
experimentation and the use of test plots. Innovation and flexibility
appear to have been important in the development of many indigenous agroforestry systems. It is part of the Haiti agroforestry
program, and is stressed by World Neighbors in its work at Cebu in
the Philippines. Experimentation allows for the continued use of
established systems while adjusting and testing for the effects of local
conditions.

A hallmark of many agroforestry systems is their superficial re-
semblance to natural succession. This imitation is an explicit part
of the work at the Xiaoliang research station in the hills of South
China. In this work an artifical broadleaved forest that resembles
natural vegetation was developed by planting and managing the
hillsides as if they were undergoing natural succession. This allowed
for demonstrable improvements in soil quality and the diversity of
soil organisms and wildlife.

Even with the use of agroforestry systems, there may be a considerable period during which timber trees produce nothing of value and are therefore ignored by subsistence farmers concerned with more immediate benefits. Some programs have addressed this problem by experimenting with incentives to farmers to care for the trees and ensure that they are not abandoned as shifting cultivators move. In Thailand, the forest villages program provides social services and helps in marketing and transportation as a way of establishing the taungya system of agroforestry. Unfortunately, the forest villages program, originally slated to be established at over two thousand sites, is languishing due to lack of consistent funding. The fate of this program highlights the difficulties of all tree-planting programs: consistent support is required through at least one complete cycle, when the first timber trees are felled. Happily for the future of tropical forests, other projects, such as those in Haiti and Ecuador, have flourished and show every prospect of continuing tree restoration for years to come.

Agroforestry and Outreach Project, Haiti

Haiti has the poorest and most thoroughly deforested countryside of any nation in the New World tropics. Viewed from an airplane over the island of Hispaniola, the boundary between Haiti and the Dominican Republic is clear: barren, eroded hills meet forest and scrub. The downward ecological spiral in Haiti has been remarkably rapid. In the late 1960s, a quarter of the country was covered by forest. Today, only seven percent of Haiti is forested, and these few remaining trees are being cut for fuelwood, charcoal and timber. If there is to be any hope of conserving the remaining wildlife and flora, trees need to be planted, woodlots established and soils stabilized. Reforestation would also stem the soil erosion which drives more and more rural people to the slums of Port au Prince.

Pioneering research by Gerald Murray and other anthropologists persuaded development planners that agroforestry projects could work in Haiti. The key was to convince farmers that tree crops were a profitable investment and that a project would support them by supplying seedlings and materials. Trees are, in fact, good income-earners: some bear fruit, some can be sold as poles, others can be converted to charcoal, and older trees may be used for lumber.

Replanting Haiti with trees from community nurseries will help stem erosion and provide income to farmers.

A farmer can make $1.50 per tree for charcoal after three to five years, and even more after the tree grows to a size which makes it useful for timber. Aside from the cash rewards, trees offer an end to erosion and soil development. In 1985, the average Haitian earned $270, while farmers could expect to make $750 from a small tree-cropping effort.

In 1981 the Pan American Development Foundation (PADF) began a long-term forestry project to help small farmers establish and protect trees on their landholdings. CARE is directing a similar effort in the northwest corner of Haiti and other forestry projects are under way in the somewhat less devastated countryside of the Dominican Republic.

The Haiti project promotes peasant ownership and stewardship of fast-growing trees that can provide fuelwood, enrich the soil and tolerate food intercropping. About forty species (such as giant leucaena and Australian pine, neem, eucalyptus and *Cassia*) are included. Native species of trees are also being planted in ever-increasing proportions—from nine percent in 1982 to thirty-four percent three years later.

Seedlings are grown in nurseries, where they are encased in small baskets of earth, light enough to be easily carried by donkeys or in headbaskets to remote sites. The planting results are analysed by PADF foresters and evaluated according to different soil, slope and planting conditions. Missionary and other private organizations broaden the project's reach on the island, while training is provided in weeding, thinning, intercropping, protection from goats and poaching, and pruning for fodder, fuel and green manure. Farmers are also taught how to construct living hedges which conserve soil by slowing run-off from Haiti's steep mountainsides.

The precondition of tree-farming, of course, is possession of a small amount of land. One aspect of the Haitian economy that favors this type of project is that most peasant families own some land. Farmers are selected if they have at least 0.2 hectares of land after food crops are planted. This amount of land is suitable for at least 250 to 500 trees, which the farmer plants as seedlings in the rainy season.

A total of some 170 private voluntary organizations (PVOs), schools and missions have accepted seedling donations, locally produced, delivered and planted with PADF advice and training. The PADF team has grown to include former Peace Corp foresters and technicians from Canada, Belgium and Haiti. The extension and outreach service is staffed by Haitian technicians and assistants. PVOs have established fairly regional nurseries, and these, too, are run by local extension workers.

It is too soon to evaluate the project's progress in reducing pressure on the scarce Haitian woodland. However, the initial statistics are impressive: during the first three years, over fifteen million trees were distributed to more than 58,000 peasants. PADF centers are now located across most of the country. Most importantly, tree planting is beginning to be seen as a money-making proposition by the farmers—a crucial step towards the economic sustainability of the program and towards successful agroforestry.

Sponsoring organizations: USAID has been the primary sponsor of the project. The project has also received support from the Canadian and Belgian government, Helvetas, Shell Limited, World Vision, Pax World Foundation, the Inter-American Foundation, and PACT (Private Agencies Collaborating Together). PADF has obtained tools and technicians from US firms and universities.

SOURCE: The Pan American Development Foundation, 1889 F Street NW, Washington DC 20006.

Growing Forest from Habitat Fragments in Guanacaste National Park, Costa Rica

The extent of deforestation varies tremendously from region to region. In some areas, tropical forests are restricted to small relict patches or galleries along streams, and overall, the number of tropical forests broken up into small fragments, rather than surviving as continuous expanses, is rapidly increasing. In these cases, the protection of woodlands may require propagating forests anew from the remaining islands of trees. Such a project is substantially more complex than establishing tree plantations on degraded lands, because it requires considerable knowledge of the former ecosystem, as well as of the factors limiting its re-establishment. The most advanced forest restoration project of this type is being conducted in the dry forests of the Pacific slope of Costa Rica. Its goal is to create the Guanacaste National Park by reforesting almost a thousand square kilometers of land around the existing core of Santa Rosa National Park.

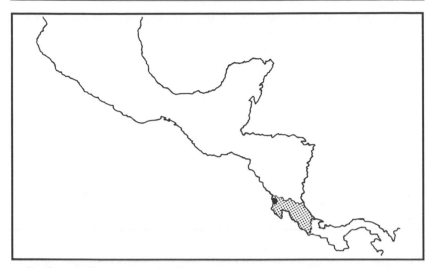

So far, deforestation in Central America has been most acute in the dry forests of the Pacific slope. All but two percent of this forest type has been cleared for farms or pasture and less than one percent of the remaining dry forest is protected from further development. Although dry forests receive less rainfall than Caribbean rainforests and suffer an extended dry season, the variation in soils and the presence of streams create microhabitats for a wide range of distinct vegetation types. Riparian forests in particular are an important dry-season refuge and help to maintain an abundance and diversity of wildlife.

Although much of the forest was initially cleared for crops such as cotton or corn, or for wood, ultimately most of it turns into pasture covered by dense stands of African grasses. These pastures are extremely susceptible to burning during the dry season, when human-ignited fires sweep across them. As a result, only fire-adapted grasses and shrubs—competitors of tree seedlings for light and nutrients—readily resprout. These two factors—fire and competition from grasses—are the major barriers to forest re-establishment in this area.

The management plan consists of purchasing land adjacent to the forested core area and employing well-established techniques of controlled burning to prevent wildfires from sweeping through the park. This is done by placing fire lanes in strategic locations through and around the park. The lanes are created by first mowing and burning narrow parallel strips of moist grasses and burning the hundred to two hundred-meter strip between these smaller strips a few months later. Throughout the fire season, constant vigilance for small spot-fires within the park is maintained.

When cattle grazing was halted in Santa Rosa National Park, in the late 1970s, it had the ironic effect of increasing the hold that the African grass had on the pastures, thereby suppressing forest regeneration through competition and fire. This observation supports park policy of using a small number of grazing animals to reduce grass growth so that the trees can compete successfully.

With the protection from fire and the judicious use of livestock, woody vegetation begins to appear spontaneously throughout the pastures. Developing a forest species composition resembling pristine dry forest, however, requires further intervention. Often the only seedlings that appear spontaneously in the pastures are those that grow from wind-dispersed seeds. Wind-dispersal is the exception, not the rule, for tropical dry forest trees, occurring in only about one-fourth of the species in the park. Furthermore, wind-dispersal is not necessarily associated with early successional species. Under these conditions, a pasture allowed to regenerate on its own may ultimately create a forest quite different from the original one. To encourage regrowth resembling naturally occurring forest, the park has planted large trees and hedgerows in pastures. These will attract seed-dispersing birds, and the seeds they drop will form the nucleus from which a forest of fruit-bearing trees can emerge. In the future park foresters will also plant seeds of native timber-producing trees, particularly at firebreaks, to create a viable agroforestry system.

The project's objectives are to freeze agricultural development in the area to be included in Guanacaste National Park, purchase lands, design a management plan for the entire park, foment pertinent research in the field and expand educational programs for both local residents and tourists. The entire project will be managed and conducted by local personnel, and will cost about $12 million to create the park, two-thirds of which is for land acquisition.

Sponsoring organizations: Financial support has come from the Mac-Arthur Foundation, The Nature Conservancy-Guanacaste Fund, W. Alton Jones Foundation, the Pew Charitable Trust, the Swedish government and the Fundacion Neotropica.

SOURCE: Daniel Janzen, Department of Biology, University of Pennsylvania, Philadelphia, Pennsylvania 19104.

Reforestation of Amazonian Bauxite Mines using Native Species, Brazil

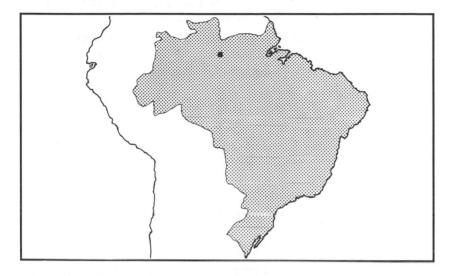

Tropical forests, particularly those of the mineral-rich Amazon Basin, are increasingly the site of mining operations. Mining is devastating to the environment. The forest is cut and cleared to reach surface deposits, and large eroded scars erupt across the landscape. Minimizing the damage caused by open-cast and strip-mining operations requires pioneering research on how to restore vegetation to these disturbed sites, as well as a long-term commitment to insure that reforestation takes hold. Such a project has been undertaken by Mineraçao Rio do Norte, a multinational company formed in the late 1970s to mine bauxite deposits in the vicinity of Trombetas in the Brazilian Amazon state of Pará. What makes this project particularly innovative is its use of a great diversity of native trees in the planting effort.

The mining is concentrated in an area of about seven thousand hectares that encompasses three large mesas covered with acidic clay soils. The process begins with the clearing of the forest with large tractors. Commercial logs are removed and the rest of the vegetation is pushed into wind-rows and burned. Debris and soil are piled up outside the mining area. The subsurface materials, or overburden, that cover the deposit are scraped off. The exposed bauxite is then loosened with explosive, excavated with heavy machinery, and hauled off by truck. The exposed trench is refilled with the overburden and levelled with large tractors. Soil stockpiles are spread over the area and the land is deep-ripped to a depth of about a meter. The result is compacted surface soil even more acidic than the original soil and now laced with nodules of bauxite, making the challenge of restoration even greater than it normally is in the poor soils of the Amazon. New mine clearings form at a rate of about seventy hectares a year—four hundred hectares in all for the years 1980 to 1987.

In the face of this ecological damage, the company decided to establish climax forest on the mine scars. This is a long-term project. More immediately, the company aimed to: revegetate the mining site with the widest possible mix of native tree species; test exotic tree species as nurse trees to help improve the soil; protect the soil from erosion; shelter slower-growing trees from wind; and test native and exotic trees as agroforestry crops during the early years of forest development.

Preparation for planting begins in the nursery, where seeds collected from the surrounding forests are stored under a variety of conditions, treated if necessary, germinated, planted in small plastic containers, and carefully tended to avoid disease and pests. Saplings are "hardened off," then transported to the mine clearings for planting. The saplings are put out in predetermined arrangements and surrounded by leguminous groundcover and nurse trees. Thereafter, the trees are tended by weeding for lianas and razor grass, eliminating leaf-cutting ants, and some replanting. To bypass the time-consuming nursery-rearing of trees, direct seeding is now being attempted.

The results so far have shown that where the saplings are replanted without replacing topsoil or ripping the overburden, growth is poor and mortality high. The trees grow best where there is topsoil and where they are accompanied by fast-growing nurse trees and leguminous groundcover. In areas of intensive treatment, unplanted forest species have spontaneously regenerated and new litter and humus layers have developed. The addition of fast-growing nurse species has also had a particularly beneficial effect on soil microclimate.

The approach is experimental. Planting conditions are varied in response to the many expected and unforeseeable problems. The latter include insect damage to saplings planted near existing forest, rapid colonization by leaf-cutting ants, toxicity of soil for some species, ignorance of dormancy-breaking requirements of certain seeds and lack of knowledge of specific growing requirements for some species. Potential problems include the blow-down of trees established in compacted soils and fires starting in the dry, exposed litter and spreading into plantations and forest. One possible barrier to the establishment of old-growth forest is that the mining method does not leave islands of trees to serve as a nucleus for the dispersal of large-seeded, slow-growing species.

Under Brazilian law, open-cast mining areas in equatorial forest must be reforested. They must also include a majority of native species. Unusually, Mineraçao Rio do Norte seriously attempts to restore natural forest. The process is expensive: the leveling of the overburden and subsequent spreading of topsoil alone costs as much as $1,500 per hectare. Growing and transplanting saplings costs another $1,000 per hectare. The possibilities of using machines, fertilizers and a direct seeding process are being explored in an attempt to reduce labor costs.

If the saplings make it through their first dry season, their chances of survival are good. In the meantime they must be protected against leaf-cutter ants, fire and other unexpected threats. The company plans to be active at the mine site for another thirty years and expects to be mining in the area for at least a century. Although no specific plan has been devised for the long-term maintenance of the planted area, the company is committed to remaining at the vanguard of environmental rehabilitation.

Sponsoring organizations: Mineraçao Rio do Norte SA and Amazon National Research Institute (consultants).

SOURCE: Oliver Henry Knowles, Environmental Adviser, Mineraçao Rio do Norte SA, Caiza Posta 23, Porto Trombetas 68275, Pará, Brazil.

Rehabilitation of Damaged Ecosystems in the Amazon Basin, Brazil

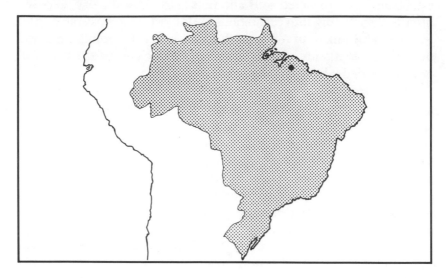

The clearing of forests for cattle pasture is one of the most important causes of deforestation in tropical America. In the Amazon, pastures made by clearing and burning large tracts are generally productive for only five to ten years. A large fraction, perhaps a fifth to a half of this cleared forest, becomes wholly degraded pasture land and can support only weeds, shrubs and exotic grasses. What factors limit forest regeneration on such lands? Answering this question is a key first step in developing ways to reforest degraded ecosytems.

In 1984, a research project based in Paragominas in the state of Pará, Brazil, began to identify the barriers to the regeneration of natural forest on pasture lands. Researchers found three main constraints, all having to do with seeds. First, the lack of seed availability is a major problem. In the research site, the pasture soil contained no tree seeds, and a mist netting study of birds in the pasture indicated that few birds had forest tree seeds or fruits in their feces (probably because species that harvest forest fruits rarely venture far out into pastures). The second problem was that of seed predation. In the experimental pasture, nine out of eleven species of seeds were removed by ants and small rodents or attacked as seedlings by leaf-cutting ants. Finally, there is a third constraint of seedling survival. Seeds face a relatively harsh environment that taxes their ability to adapt physiologically. Plants in the pasture, as opposed to those in the forest, encounter lower soil moisture, higher rates of evapotranspiration, and heavier doses of solar radiation. In the Amazon, soil moisture stress and low atmospheric water pressure are particularly intense during the five-month dry season; this is precisely the time, according to preliminary observations, when forest tree seedlings, especially of small-seeded colonizing species, are apt not to survive.

The next phase of the research will compare the survival rates of seedlings planted in forests to those of seedlings placed in pasture. The different species will also be compared for survival, growth, and drought tolerance or avoidance. Through identifying the important traits that enable certain tree species to survive in pasture, this study will help in selecting tree species that have the best chance of succeeding in reclaiming unproductive tropical lands. Even then, as the research results suggest, reclamation will be exceedingly difficult. In the final analysis, the most effective way of maintaining forest cover and preventing land degradation is to conserve existing forests.

Sponsoring organizations: Funds provided by the National Geographic Society and the National Science Foundation; help and cooperation from EMBRAPA, Centro de Pesquisa Agropecuaria do Tropico Humido, Belem, Pará, Brazil.

SOURCE: Christopher Uhl, 202 Buckhout Laboratory, Pennsylvania State University, University Park, Pennsylvania 16802.

Plan Bosque: Incentives for Planting and Tending Trees, Ecuador

Each year, Ecuador loses more than 300,000 hectares of its natural forests. This forested nation now spends up to $90 million a year to import pulp and paper products. Wood shortages are acute over much of the Andes, where dry conditions limit forest growth. But even the vast sea of forest found east of the Andes in the country's Oriente province is rapidly being cleared and developed.

Regrowing forest on cleared lands will alleviate some of the pressure. Reforestation will also have the less tangible effect of instilling a greater "forest consciousness" in the country's people. Yet despite

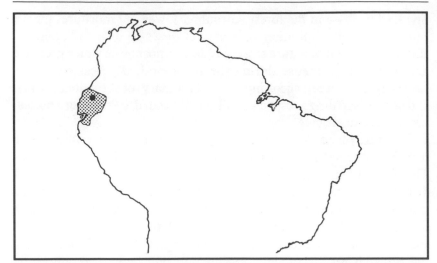

the obvious long-term advantages of reforestation, a fundamental question remains: how can people be motivated to invest time and labor in caring for trees when there may not be any benefit for years? For peasants who are struggling to survive day to day, this is indeed the issue. They cannot afford the luxury of supplying trees with scarce water and land while their children lack food; when their plots of land cease to be productive, they simply move on to more distant plots, often in previously forested areas.

In response to this situation, Ecuador established a tree fund, "Plan Bosque," in 1985. Derived from a tax on oil export revenues, and beginning with the equivalent of $12 million, it aimed to provide incentives to peasants to plant and care for trees. The need for the fund was obvious: until the year of its founding, a mere 3,000 hectares were replanted annually. Plan Bosque's original targets were the arid and mountainous regions where wood shortages are critical. Recently, however, it has expanded into the Oriente (Amazon) region, where 1,800 hectares in sixty holdings were planted in 1986.

Plan Bosque operates by providing loans. These are made to individuals, cooperatives, tribal organizations and companies; the major requirement is that the applicants own land in sufficiently good condition to support successful tree planting. When the funding agency, the Direccion Nacional Forestal (DINAF), receives an application, a representative visits the proposed reforestation site and judges its suitability. If approved, funds are authorized for seed, soil preparation, planting, maintenance, insurance and extension service for two years. Upon signing the contract, a bank provides the actual cash disbursement to the landowner, who must spend it within two years.

At the end of the two-year period, representatives of DINAF inspect tree-planting sites to verify that survivorship is greater than the threshold set for each species, which ranges from seventy to ninety percent. Both exotic species (such as eucalyptus, pine, cypress and leucaena) and native species (such as cedro and amarillo) are planted. If a planting project has been successful, DINAF pays off the bank loan and interest with funds from the oil export tax. However, if seedlings died through negligence, it becomes the responsibility of the planter to pay off the loan with full interest (which is thirty-six percent in Ecuador) and penalties. In ten to twenty years, after the first wood harvest, the planter repays the government the amount of the loan, but not the interest.

In 1986 Plan Bosque funded 590 projects covering 29,000 hectares. While the total area planted is vast, the average project site, at fifty hectares, is quite small. In fact, projects must be no more than a hundred hectares to qualify for loans. The costs are low, too, about $250 per hectare, although the total cost through 1987 is over $6 million. The original investment will not be recovered for at least ten years, but Plan Bosque is potentially financially self-supporting.

Plan Bosque is the result of four years of organizational and lobbying efforts by Fundacion Natura, an Ecuadorian conservation organization. The immediate benefit of the plan is the tremendous increase in tree plantings on barren lands. The long-term benefit should be a greater commitment on the part of the Ecuadorian government to conservation of forests and hence to the preservation of significant tracts of the country's remaining forest lands.

Sponsoring organizations: Direccion Nacional Forestal and Fundacion Natura.

SOURCE: Roque Sevilla, former Director, National Forestry Program, Ministry of Agriculture, Quito, Ecuador. Present address: PO Box 243, Quito, Ecuador.

Xiaoliang Water and Soil Conservation Project, China

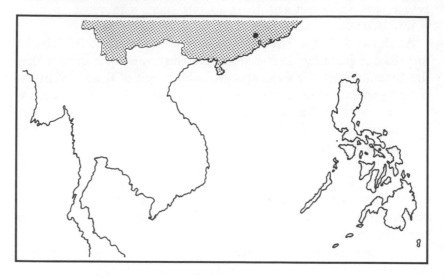

The area around Xiaoliang in Guangdong Province, where lateritic red soil once supported rich evergreen monsoon forests, is typical of barren, eroded lands found in China today. Unlike most currently deforested areas, Xiaoliang's ecological crisis is the product of centuries-old colonization and resource exploitation. Over time, deforestation greatly affected the local hydrological cycle, leaving the farmers at the mercy of drought and floods. In 1959, the South China Institute of Botany, Academia Sinica, and the Xiaoliang Province Station of Water and Soil Conservation began to study both natural and artificially generated revegetation of the eroded hillsides.

Due to the virtually complete removal of evergreen tropical forest, the annual erosion, which has occurred over the past hundred years, has reached a centimeter or more a year. Soil humus has declined from four to 0.6 percent, the climate has become drier and the water cycle one of extremes. With the goal of reversing some of this environmental degradation, a program of revegetation that replicates some of the qualities of natural forest succession has begun. The developmental stages are as follows:

- *Pioneer arboraceous community*: Fast-growing, drought resistant trees such as pines, eucalyptus and acacias are planted.
- *Broadleaved, mixed forest*: Leguminous plants form the canopy of shade-tolerant trees, shrubs and palms.

- *Economic activities*: agroforestry and animal husbandry integrated into the system.

After twenty-five years of effort, the restored ecosystem covers 433 hectares. Researchers have monitored intensive plots for climate, soil, and hydrology since 1980. These results have been detected:

Biological effects. There has been a general increase in the diversity of organisms of different broad taxonomic classes. Before 1973, most of the experimental area was covered by single species stands of *Pinus massoniana* and *Eucalyptus exserta*. With the resurgence of the mixed forest, total plant species increased from ten to 320. The number of birds found in the mixed forest was 144 per hectare in the broadleaved forest versus 22 and 93 respectively in the eucalypt forest and barren hills. There are five times more soil animals in the eucalypt forest than in barren hills but an astonishing fifty times more soil animals in the broadleaved forest. Soil microbes increased as well.

Microclimate effects. The annual fluctuation in soil temperature decreased from 14.3°C to 13°C and the relative humidity increased from 83 to 87 percent. Organic matter in the soil doubled to 1.1 percent and it became less acid.

Hydrological and erosion effects. The water-carrying capacity of the surface soil increased and the erosion of soil declined from nearly a ton to three kilograms per hectare per year. The water-holding capacity of the subsoil was highest on the eroded hills, presumably because water was taken up by the roots of trees and passed off in evapotranspiration.

In addition to these ecological effects, the project has had an important social and economic impact. The conservation station has moved from being dependent on government funds to generating income through its agroforestry activities. For the wider community, the project has reduced the amount of eroded mud washing into the lands of neighboring villages, and rice production has increased seven-fold through the control of soil erosion and flooding. The project has also served as a model and provided an extension service for agroforestry work in Guangdong Province.

Sponsoring organizations: South China Institute of Botany, Academia

Sinica, Guangdong Institute of Entomology, Guangsho Institute of Geography, Xiaoliang Water and Soil Conservation Experiment and Extension Station.

SOURCE: Yu Zuo-Yue, South China Institute of Botany, Academia Sinica, Guangzhou, China.

The Forest Villages, Thailand

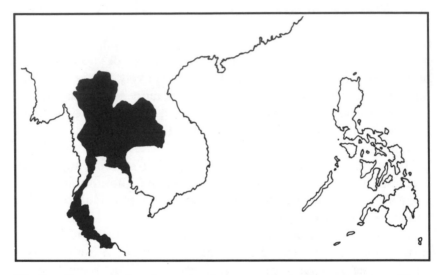

Thailand's remaining forest lands are under intense pressure. Since 1970, the number of people engaging in shifting agriculture has rocketed from 300,000 to over 700,000. In 1980, perhaps 800,000 hectares of forest land were under shifting agriculture, while another 400,000 hectares were cleared. One possible solution was to cut back on the level of "shifting" by offering farmers and their families some measure of economic security.

The forest village project seeks to do just that. Started by the Forest Industries Organization (FIO), the Royal Thai Forest Department and the Royal Thai Family, the project provides a mixture of economic and social incentives to farmers to settle in degraded forest areas throughout the country. Its title comes from the fact that it actually sets up villages. There, the government provides social services and cash payments to promote agroforestry. Importantly, the agroforestry system that is being used is the taungya approach—

one that makes sense to the villagers and is most appropriate to the Thai countryside. The taungya system was originally introduced in British colonies to encourage the planting of economically valuable trees. Through this network of villages—originally intended to number two thousand—the project hopes to reduce the clearing of forest, improve villagers' living standards and encourage reforestation.

In any given village, a family is allocated 1.6 hectares a year to clear for crops and plant trees, and an additional 0.16 hectares for a kitchen garden and home. Each site is cropped for three years, so families farm 4.8 hectares at a time. With around a hundred families in each village, this means that 150 hectares will be used on a sustained, and ecologically sound, basis.

The FIO provides other services to the villages as well, including medical care, drinking water and electricity, primary schools and textbooks. Monetary incentives for the establishment of tree plantations include cash for clearing, planting and two weedings of allocated land. Another payment is provided for tree survival through the three years of cropping and additional non-monetary incentives —free transportation, for instance—are provided for continued work.

In the taungya system, which is a linchpin of the project, both subsistence and cash crops are grown. In the forest clearings the major crops are rice, corn, sesame, sweet potatoes and cassava. The homegardens boast maize, cassava, pumpkins, chili peppers, various beans, tubers, squashes, millets and some medicinal plants and grain. Teak, eucalyptus and melia are most commonly planted for wood; fruit and nut trees are grown as well. In some areas rubber trees are grown and cash from the sale of latex is divided between the FIO and the forest farmers, with the farmers claiming seventy percent of the profits. With its thirty percent, the FIO buys fertilizer and tools for the villagers. Trees are planted in degraded forest located in forest reserves or other government land, and crops are planted between rows. After three years of cultivation of a site, a family cultivates a new area of degraded forest for crops and continues to tend the trees at the old site. The trees will provide agroforestry crops of the future.

Originally it was hoped that the program could establish two thousand villages on 32,000 hectares of degraded forest land. In 1981 there were twenty-six villages; by 1987 there were only thirty-five. Nonetheless, families have benefited: the average family earned about $260 in crop sales in 1981; with added financial incentives and wages, this figure came to about $700—a total far superior to the average Thai income. Unfortunately, government funding was withdrawn before a complete rotation of the first important cash tree-crops, eucalyptus and teak. Still, the organizers feel that the project is sound, and that, with the continuing support it receives from participants and private organizations, it will survive. Certainly the forest villages show that it is possible to improve peoples' living conditions while reducing the need for further destruction of natural resources.

Sponsoring organizations: Forest Industry Organization and the Royal Forestry Department.

SOURCE: S.A. Boonkird, E.C.M. Fernandes and P.K.R. Nair, Forestry Faculty, Kasetsart University, Bangkok, 10903, Thailand, and International Council for Research in Agroforestry, P.O. Box 30677, Nairobi, Kenya.

Soil Conservation on Steep Tropical Slopes, Philippines

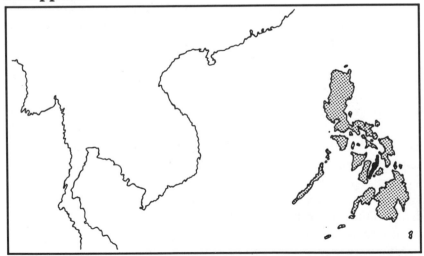

Farming on mountainous tropical slopes can rapidly lead to soil erosion. This leads to flooding and siltation of water courses, and to soil impoverishment which forces farmers to abandon land and move to agricultural frontiers, so perpetrating a descending spiral of land degradation. Throughout the tropics, forests recede up mountain slopes, are replaced temporarily by poor farm plots for a few years, and ultimately become denuded hillsides covered with tough, fire-resistant grasses. Any comprehensive conservation effort will have to tackle this problem and its underlying causes. Many projects, in fact, are attempting to stabilize land use on tropical hillsides through the introduction of agroforestry. One exemplary project was undertaken in 1982 by World Neighbors on the island of Cebu in the Philippines.

The Cebu project attempts to stop soil erosion and increase the productivity of cultivated slopes by introducing simple conservation measures. World Neighbors is also actively seeking land titles for small farmers participating in the conservation program. The principle is that farmers who have title to their land will have an interest in its long-term viability.

In 1981–82 specific sites for extension work were determined, a profile of conservation measures suited to local conditions developed and a training program was established. Conservation techniques were tested on small plots and demonstrated to program participants. These included: the construction of contour ditches planted with leucaena to control runoff and provide forage, the use of compost, and intercropping with legumes. Local farmers who were particularly enthusiastic about the program were selected to spearhead extension efforts.

Two primary sites were identified, and in discussions with the project director, local villagers voiced a litany of seemingly insurmountable farming problems. However, many of the problems centered on poor productivity, largely the result of soil erosion and nutrient loss. Considerable support developed for a communal labor group that would spend one day a week working on soil conservation projects. This cooperative effort was not difficult to arrange because work-sharing is a common feature of village life in the Philippine countryside. A series of seminars was given on basic contouring and other soil conservation techniques. At first, the World Neighbors project director led these sessions, but eventually knowledgeable villagers familiar with the techniques took over the seminar presentations. The major capital investment for the seminars was a slide projector used to show villagers the progress others were making on their farms.

On the eroded hillsides, World Neighbors introduced a number of innovations: the construction of rubble walls, hedgerows, contour canals, gully dams, drainage canals and simple composting systems.

Villagers were also taught how to construct a simple A-frame level for setting rows and canals along contours. Contour walls were made from rubble piled on slopes with leucaena planted below the wall and forage grass planted above it, while the hedgerows consisted of dense planting of napier grass, leucaena and Madre de Cacao (*Gliricidia*). (Diversification from the sole use of leucaena has been necessary due to a psyllid insect infestation on leucaena; *Calliandra* and *Gliricidia* are two promising alternatives.) The planting of forage has

Simple, inexpensive techniques are used to determine planting patterns for hillside farming that will help conserve soil in the Philippines. Based on illustrations from *Soil Conservation on Steep Tropical Slopes*.

allowed farmers to keep goats, providing an immediate economic return for their conservation efforts.

Farmers were also taught new techniques in pest control and horticulture. They now grow greater quantities and varieties of vegetables than before. The program emphasizes continued experimentation with new plants and planting techniques. Farmers are encouraged, for example, to set up small test plots.

After two years, seventy-four farmers had been involved in the project, and the equivalent of twenty-five kilometers of erosion-control measures had been built. By the end of 1987 three new sites had been added, and project workers now expect 750 farmers to adopt the soil and water conservation techniques. The project's focus has broadened to encompass soil fertility management and simple techniques to increase the productivity of reclaimed soils, which are being publicized at seminars and through a booklet. In Cebu, the seed has been sown but, as in most rural development programs, it will be years before the program proves itself. The final test, as always, is whether farmers can produce enough for their subsistence using lasting, sustainable agriculture.

Sponsoring organization: World Neighbors Soil and Conservation Project, PO 286, Cebu City, Philippines.

SOURCE: Robert Curtis, World Neighbors, 5116 Portland Avenue, Oklahoma City, Oklahoma 73112.

Village Forest Project, Uganda

In many areas, the production of fuelwood on degraded lands is critical for the protection of remaining intact forests. Since 1984, CARE and USAID have co-sponsored a project with the government of Uganda which involves planting trees for fuelwood and the development of agroforestry. A major goal of this project is to reduce pressure on the Kibale Forest Reserve, which protects an area of lowland, humid tropical forest in southwestern Uganda.

One force behind the creation of the Village Forest Project was the New York Zoological Society's effort to preserve the Kibale Forest Reserve by improving the living conditions of local people. A second impetus came from the CARE-USAID matching grants program,

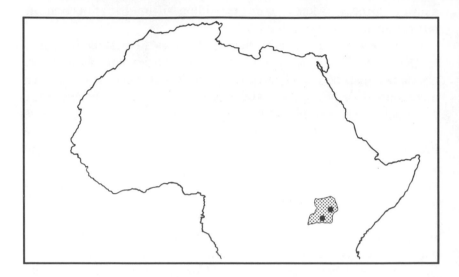

designed to fund new projects in the area of agroforestry. By 1983 it was clear that deforestation rates in Uganda were unacceptably high and that soil infertility and shortage of fuelwood were major national problems. The Ugandan Forestry Department was relatively weak, doing little to promote conservation and reforestation despite a strong tree-planting ethic among the population. One goal of the Village Forest Project, accordingly, was to strengthen the Forestry Department in the areas of forestry extension, reforestation, agroforestry and soil conservation. The other objectives were reducing deforestation, increasing farmers' income from wood sales, and mitigating shortages in the fuelwood supply. The project has kept the costs of producing and planting trees low by establishing a widespread network of small, community managed nurseries. These nurseries attract more local participation, especially of women, and foster on-farm tree planting.

As of 1987, the project was operating in eleven districts in both the eastern and western ends of the country and supporting 274 nurseries producing about three million seedlings a year. The seedlings were primarily eucalyptus and cypress for woodlots and a smaller number of agroforestry species, such as *Cassia siama*, *Casuarina equisetifolia*, *Sesbania sesban*, and *Markhamia platicalyx*. The project started by using bare-root stock of eucalyptus and cypress. This helped build a good working relationship between CARE and the Forestry Department, since these trees are more easily managed than agroforestry species and provide immediate benefits in terms of poles, fuel and shelter.

Ultimately, the agroforestry systems introduced in 1986 will benefit the small landholders more than tree crops alone, because its methods produce a diversity of wood and horticultural products that can be used domestically or sold profitably in small quantities. Other agroforestry techniques such as alley-cropping, boundary planting and intercropping will help restore soil fertility. The project is also promoting the establishment of self-sustaining nurseries and extension programs and seminars, provided by a highly mobile forestry staff.

Participating farmers have been strongly supportive of the project, as it answered their need for both wood and an improved standard of living. They consider trees to have economic value and have been willing to pay for seedlings, although farmers were skeptical about the use of bare-root stock. Farmers in the agroforestry programs were generally enthusiastic about future possibilities, especially with the woodlot trees eucalyptus and cypress. There appears to have been an increase in the number of woodlots and planting, although as of yet there has been no reduction in the amount of incursion into natural forest reserves. Although these successes appear modest, they have been achieved during a period of tremendous political instability in Uganda. If the farmers' responses are any indication, Ugandan agroforestry will have a long and productive future.

Sponsoring organizations: CARE, USAID and the Ugandan Forestry Department.

SOURCE: Charles Tapp, CARE, 660 First Avenue, New York, New York, 10016.

NOTES

PREFACE

1. Henry Mayhew, *London Labour and London Poor* (1851).
2. T. Erwin, "Beetles and Other Insects of Tropical Forest Canopies at Manaus, Brazil, Sampled by Insecticidal Fogging," in *Tropical Rain Forest: Ecology and Management*, ed. S.L. Sutton, T.C. Whitmore and A.C. Chadwick (Oxford: Blackwell Scientific, 1983).
3. I. Rubinoff, "Tropical Forests: Can We Afford Not to Give Them a Future?" in *The Ecologist* 12 (6) (1984): 253–258.
4. R. Rhodes, *The Making of the Atomic Bomb* (New York: Touchstone, Simon & Schuster, 1988).

PART I   THE SPECTER OF DEFORESTATION

1. C.F. Bennett, "Human Influences on the Zoogeography of Panama," in *Ibero-American* 5 (1968): 1–112; A. Gomez-Pompa, "On Maya Silviculture," in *Journal of Mexican Studies* (1987).
2. C. Uhl, "You Can Keep a Good Forest Down," in *Natural History* April 1, 1984: 71–79; National Academy of Sciences, *Ecological Aspects of Development in the Humid Tropics* (Washington, DC: NRC/NAS Press, 1982).
3. Uhl, "You Can Keep a Good Forest Down," op. cit.
4. H.S. Lee, "Silvicultural Management in Sarawak," in *Malaysian Forester* 45 (1982): 485–496.
5. T. Whitmore, *Rainforests of the Far East* (Oxford University Press, 1975).
6. Lee, "Silvicultural Management in Sarawak", op. cit.; P.P.O. Kio and S.A. Ekwebelam, "Plantation versus Natural Forests for Meeting Nigeria's Wood Needs," in *Natural Management of Tropical Moist Forests: Silvicultural and Management Prospects of Sustained Utilization*, ed. F. Magen and J.R. Vincent (New Haven: Yale School of Forestry, 1987).
7. H.S. Lee, "The Development of Silvicultural Systems in the Hill Forests of Malaysia," in *Malaysian Forester* 45 (1982): 1–8; I.D. Hut-

chinson, "Improvement Thinning in Natural Tropical Forests: Aspects and Institutionalization," in *Natural Management of Tropical Moist Forests*, ed. Magen and Vincent, op. cit.

8. Office of Technology Assessment, *Technologies to Sustain Tropical Forest Resources* (Washington, DC: OTA, 1984).

9. National Academy of Sciences, *Ecological Aspects of Development in the Humid Tropics*, op. cit.

10. P.A. Sanchez, D.E. Bandy, J.H. Villachica and J.J. Nicholaides, "Amazon Basin Soils: Management for Continuous Crop Production," in *Science* 216 (1982): 821–827.

11. F. Lappé and J. Collins, *Food First* (New York: Ballantine Books, 1978); E. de G.R. Hansen, "Let Them Eat Rice," in *Bordering on Trouble: Resources and Politics in Latin America* (Washington, DC: World Resources Institute, 1986).

12. P.C. Matteson, M.A. Altieri and W.C. Gagne, "Modification of Small Farmer Practices for Better Pest Management," in *Annual Review of Entomology* 29 (1984): 383–402.

13. World Resources Institute and International Institute for Environment and Development, *World Resources 1986* (New York: Basic Books, 1986).

14. J.P. Lanly, *Tropical Forest Resources* (Rome: FAO, 1982); A. Grainger, "Quantifying Changes in Forest Cover in the Humid Tropics: Overcoming Current Limitations," in *Journal of World Forest Resources Management* 1 (1984): 3–63.

15. P.M. Fearnside, "Spatial Concentration of Deforestation in the Brazilian Amazon," in *Ambio* 15 (2) (1986): 72–79.

16. P.M. Fearnside, "Deforestation and International Economic Development Projects in Brazilian Amazonia," in *Conservation Biology* 1 (1987): 214–221.

17. F. Nectoux and N. Dudley, *A Hardwood Story: An Investigation into the European Influence in Tropical Forest Loss* (London: Friends of the Earth, 1987).

18. World Resources Institute, *Tropical Forestry Action Plan* (Washington, DC: 1985).

19. Nectoux and Dudley, *A Hardwood Story*, op. cit.

20. R. Repetto, "Creating Incentives for Sustainable Forest Development," in *Ambio* 16 (1987): 94–99.

21. A. Grainger, "The State of Tropical Rainforests," in *The Ecologist* 10 (1980): 6–54; Repetto, "Creating Incentives," op. cit.

22. W. Denevan, *Causes of Deforestation and Forest Woodland Degradation in Tropical Latin America* (Washington, DC: Office of Technology Assessment, 1982).

23. Lanly, *Tropical Forest Resources*, op. cit.; N. Myers, *Conversion of Tropical Moist Forests* (Washington, DC: National Academy of Sciences, 1980).

24. V. Plumwood and R. Routley, "World Rainforest Destruction: The Social Factors," in *The Ecologist* 12 (1982): 4–22.

25. Denevan, *Causes of Deforestation*, op. cit.; J.E. Spencer, *Shifting Cultivation in Southeast Asia* (Berkeley: University of California Press, 1977).

26. B. Rich, "Multilateral Development Banks: Their Role in Destroying the Global Environment," in *The Ecologist* 15 (1985): 56–68.

27. N.J.H. Smith, *Rainforest Corridors: The Transamazon Colonization Scheme* (Berkeley: University of California Press, 1982).

28. Fearnside, "Spatial Concentration of Deforestation," op. cit.; Rich, "Multilateral Development Banks," op. cit.

29. C. Caufield, *In the Rainforest* (New York: Alfred A. Knopf, 1985); World Resources Institute, *World Resources 1986*.

30. T. Barry and D. Preusch, *The Central America Fact Book* (New York: Grove Press, 1986).

31. Lappé and Collins, *Food First*, op. cit.

32. F. Samana, "Women in Agriculture," in *Subsistence Agriculture Improvement Manual*, Wau Ecology Institute publication no. 10, 1986.

33. J. Nations and D. Komer, *Rainforest, Cattle and the Hamburger Society* (Austin: Center for Human Ecology, 1982); S. Hecht, "Environment, Development and Politics: Capital Accumulation in the Livestock Sector in Eastern Amazonia," in *World Development* 13 (1985): 663–684.

34. Hansen, "Let Them Eat Rice," op. cit.

35. M.J. Dourojeanni, *Renewable Natural Resources of Latin America and the Caribbean: Situations and Trends* (Washington, DC: World Wildlife Fund-US, 1982).

36. P. Blaikie, *The Political Economy of Soil Erosion in Developing Countries* (London: Longman, 1985).

37. L.R. Brown, "A Generation of Deficits," in *State of the World 1986* (Washington, DC: Worldwatch Institute, 1986).

38. World Bank, *World Development Report* (Oxford University Press, 1985).

39 IUCN, *Vietnam National Conservation Strategy* (New Delhi: IUCN, 1987).

40. C. Mackie, "The Lessons Behind East Kalimantan's Forest Fires," in *Borneo Research Bulletin* 16 (2) (1984): 63–74.

41. Lanly, *Tropical Forest Resources*, op. cit.

42. Fearnside, "Spatial Concentration of Deforestation," op. cit.

43. Plumwood and Routley, "World Rainforest Destruction," op. cit.

44. B. Neitschmann, "Indonesia and Bangladesh: Economic Development by Invasion of Indigenous Nations," in *Cultural Survival* 10 (2) (1986): 2–12.

45. S. Davis, *Victims of the Miracle* (Cambridge University Press, 1977).

46. Hansen, "Let Them Eat Rice," op. cit.

47. UNESCO/UNEP/FAO, *Tropical Forest Ecosystems* (1978).

48. L.C. Brown and E.C. Wolf, *Soil Erosion: Quiet Crisis in the World Economy*, Worldwatch paper no. 60, 1984.

49. A. Wijkman and L. Timberlake, *Natural Disasters: Acts of God or Acts of Man* (London: Earthscan, 1984).

50. E. Salati, T.E. Lovejoy and P.B. Vosse, "Precipitation and Water Recycling in Tropical Rainforests," in *The Environmentalist* 3 (1983): 67–72.

51. E.O. Willis, "Populations and Local Extinctions of Birds on Barro Colorado Island, Panama," in *Ecological Monographs* 44 (1974): 153–169.

52. S.P. Hubbell and R.B. Foster, "Diversity of Canopy Trees in a Neotropical Forest and Implications for Conservation," in *Tropical Rain Forest: Ecology and Management*, ed. S.L. Sutton, T.C Whitmore and A.C. Chadwick (Oxford: Blackwell Scientific, 1983).

53. E.C. Wolf, *On the Brink of Extinction: Conserving the Diversity of Life*, Worldwatch Paper no. 78, 1987.

54. L. Lovejoy, "A Projection of Species Extinction," in *The Global 2000 Report* 2 (1982): 328–331; P.R. Ehrlich and A.H. Ehrlich, *Extinctions* (New York: Random House, 1982).

55. N. Myers, "Tropical Deforestation and Species Extinctions: the Latest News," in *Futures* 17 (1985): 451–463.

56. N.R. Farnsworth, "The Potential Consequences of Plant Extinction in the United States on the Current and Future Availability of Prescription Drugs," in *The First Resource, Wild Species in the American Economy*, ed. C. Prescott-Allen and R. Prescott-Allen (New Haven: Yale University Press, 1982).

57. National Academy of Sciences, *Ecological Aspects of Development in the Humid Tropics*, op. cit.

PART II: The Case Studies

Chapter 1: Forest Reserves

The Community Baboon Sanctuary: An Approach to the Conservation of Private Lands

Horwich, R.H., "The Geographic Distribution of the Black Howler Monkey (*Alouatta pigra*) in Central America and Efforts to Conserve It in Belize," in *Primatologia en Mexico: Comportamiento, Ecologia, Aprovechamiento y Conservacion de Primates No Humanos en Mexico*, ed. A. Estrada (Mexico City: UNAM 1987).

Horwich, R.H., "A Community Baboon Sanctuary in Belize," in *Primate Conservation* 7: 15.

Horwich, R.H., and E.D. Johnson, "Geographical Distribution of the Black Howler (*Alouatta pigra*) in Central America," in *Primates* 27 (1986): 53–62.

Horwich, R.H., and J. Lyon, "Development of the Community Baboon Sanctuary in Belize: An Experiment in Grassroots Conservation," in *Primate Conservation* (1987).

La Amistad Biosphere Reserve
Torres, H., and L. Hurtado (eds), *Parque Internacional de La Amistad— Plan General de Manejo y Desarrollo* (Madrid: Instituto de Cooperacion Iberoamericano, 1987).
Torres, H., and L. Hurtado (eds), *Reserva la Biosfera de La Amistad: Una Estrategia para su Conservacion y Desarrollo* (Turriabla, Costa Rica: Tropical Agricultural Research and Training Center, 1987).

The Kuna Yala Biosphere Comarca
Houseal, B., C. MacFarland, A. Archibold and A. Chiari, "Indigenous Cultures and Protected Areas in Central America," in *Cultural Survival Quarterly* 9 (1).

A Bi-National Approach to the Protection of Indian Lands
MacDonald, T., Jr, "Anticipating *Colonos* and Cattle in Ecuador and Colombia," in *Cultural Survival Quarterly* 10 (1986): 33–36.

Manu Biosphere Reserve
Munn, C.A., "Ciencia y Tourismo en la Reserva de Biosfera del Manu," unpublished paper available from the Conservation Association for the Southern Rainforests of Peru, Cuzco, Peru.

Protection and Development in and about Khao Yai Park
Brockelman, W.Y., "Nature Conservation," in *Thailand Natural Resources Profile*, ed. A. Arbhabhirama, D. Phantumvanit, J. Elkington and P. Ingkasuwan (Bangkok: Thailand Development Research Institute, 1987).
Dobias, R.J., *WWF/IUCN Project 3001: Elephant Conservation and Protected Area Management*. Final report to WWF, IUCN and the Royal Thai Forest Department, 1985.
Dobias, R.J., and C. Khontong, "Integrating Conservation and Rural Development in Thailand," in *Tigerpaper* Vol. XIII, No. 4. (Bangkok: FAO, 1986).
Suvanakorn, P., and R.J. Dobias, "Using Economic Incentives to Improve Park Protection," in *Conservation of Tropical Forests: Case Studies of Approaches that Have Worked* (Gland: IUCN in press).

Protecting Wildlife and Watersheds at Dumoga Bone
Sumardja, E., Tarmudji and J. Wind, *Nature Conservation and Rice Production in the Dumoga Valley, Dumoga Bone National Park, North Sulawesi*. Proceedings of the Third World Conference on National Parks, Bali, Indonesia, 1982.

Wind, J., and E. Sumardja, *Dumoga-Indonesia: World Bank Project for Irrigation and Water Catchment Protection*. Conference on Sustainable Development, IIED, London, 1987.

Korup National Park
Gartlan, S., *The Korup Regional Management Plan: Conservation and Development in the Ndian Division of Cameroon* (Gland: World Wildlife Fund–US, 1985).
Cloutier, A., and A. Dufresne, *Plan de Gestion, Parc National de Korup* (Quebec: Parcs Canada, 1986).
Gartlan, S., and H. Macleod (eds), *Proceedings of the Workshop on Korup National Park, Mundemba, Ndian Division, South-West Province, Cameroon* (Gland: WWF/IUCN, 1986).
Gartlan, S., and D. Momo, "Korup: New Approach to Conservation," in *IUCN Bulletin* 17 (1986): 1–3.

CHAPTER 2: SUSTAINABLE AGRICULTURE

Lessons from Mayan Agriculture
Gomez-Pompa, A., "On Mayan Silviculture," in *Journal of Mexican Studies* (Berkeley: University of California Press, 1987).
Gleissman, S.R., R. Garcia and M. Amador, "The Ecological Basis for the Application of Traditional Agriculture Technology in the Management of Tropical Ecosystems," in *Agro-ecosystems* 7 (1981): 173–185.

Iguana Ranching: A Model for Reforestation
Chapin, M., "The Panamanian Iguana Renaissance," in *Grassroots Development* 10 (2) (1986): 2–7.
Fitch, H., R. Henderson and D. Hillis, "Exploitation of Iguanas in Central America," in *Iguanas of the World: Their Behavior, Ecology and Conservation,* ed. G. Burghardt and A. Rand (Park Ridge, NJ: Noyes Publications, 1982).
Henderson, R., "Aspects of the Ecology of the Juvenile Common Iguana *(Iguana iguana)*," in *Herpetologica* 30 (1974): 327–332.
Hirth, H., "Some Aspects of the Natural History of *Iguana iguana* on a Tropical Strand," in *Ecology* 44 (3) (1963): 613–615.
Werner, D., and A. Rand, "Manejo de la Iguana Verde en Panama," in *Symposio de Conservacion Manejo Fauna Silvestre Neotropical* (Arequipa, Peru: Fourth Latin American Zoological Congress, 1986).
Werner, D., "Research on Management of an Endangered Species in Panama: The Green Iguana," in *Biological Conservation Newsletter* 21 (1984): 1–2.
Werner, D., "Iguana Management in Central America," in *BOSTID Developments* 6 (1) (1986): 1–4–6.

Resource Management by the Kayapó
Posey, D., "Indigenous Management of Tropical Forest Ecosystems: The Case of the Kayapó Indians of the Brazilian Amazon," in *Agroforestry Systems* 3 (1985): 139–158.
Linares, O., "Garden Hunting in the American Tropics," in *Human Ecology* 4 (4) (1976): 331–349.
Smith, N., "Human Exploitation of Terra Firma Fauna in Amazonia," in *Ciencia e Cultural* 30 (1) (1977): 17–23.

Japanese Farming in the Amazon Basin
Staniford, P., *Pioneers of the Tropics: The Political Organization of Japanese in an Immigrant Community in Brazil*, London School of Economics Monographs on Social Anthropology no. 45 (London: Athlone Press, 1973).
Flohrschutz, G.H.H., *Analise Economica de Establecimentos Rurais do Municipio de Tome-Acu, Pará—Um Estudio de Caso* (Belém, Brazil: Empresa Brasiliera de Pesquisa Agropecuaria, Centro de Pesquida Agropecuaria, Centro de Pesquisa Agropecuaria do Tropico Humido, documento 19, 1983).

Long-term Management of Swiddom-Fallows by Bora Indians
Denevan, W.M., J.M. Treacy, J.B. Alcorn, C. Padoch, J. Denslow and S.F. Paitan, "Indigenous Agroforestry in the Peruvian Amazon: Bora Indian Management of Swidden Fallows," in *Interciencia* 9 (1984): 346–357.
Denevan, W.M., and C. Padoch (eds), "Swidden-fallow Agroforestry in the Peruvian Amazon", *Advances in Economic Botany* (New York: New York Botanical Garden, in press).

Market-oriented Agroforestry in the Amazon
Padoch, C., J. Chota Inuma, W. de Jong and J. Unruh, "Amazonian Agroforestry: A Market-oriented System in Peru," in *Agroforestry Systems* 3 (1985): 47–58.

Javanese Home Gardens
Fernandes, E., and P. Nair, "An Evaluation of the Structure and Function of Tropical Homegardens," in *Agricultural Systems* 21 (1986): 1–14.
Michon, G., "Village Forest Gardens in West Java," in *Plant Research and Agroforestry*, ed. P. Huxley (Nairobi: ICRAF, 1983).
Soemarwoto, O., and I. Soemarwoto, "The Javanese Rural Ecosystem," in *An Introduction to Human Ecology Research on Agricultural Systems in Southeast Asia*, ed. T. Rambo and E. Sajise (Los Banos: University of the Philippines, 1984).

An Extension Service for Stabilizing Shifting Agriculturalists
Gagne, W.C., and J.L. Gressitt, "Conservation in New Guinea," *Biogeography of New Guinea*, ed. J.L. Gressitt. *Monographaea Biologicae* 42 (1982): 945–966.
Gagne, W., "Stabilizing Shifting Agriculture," in *Harvest* 6 (1980): 192.
Gueltenboth, F., *Subsistence Agriculture Improvement Manual*. Wau Ecology Institute Handbook No. 10 (1984).

CHAPTER 3: NATURAL FOREST MANAGEMENT

A Sustainable Silviculture System for Forests in Suriname
Boxman, O., N.R. de Graaf, J. Hendrison, W.B.J. Jonkers, R.L.H. Poels, P. Schmidt and R. Tjon Lim Sang, "Towards Sustained Timber Production from Tropical Rain Forests in Suriname," in *Netherlands Journal of Agricultural Science* 33 (1985): 125–132.
de Graaf, N.R., *A Silvicultural System for Natural Regeneration of Tropical Rain Forest in Suriname* (Agricultural University of Wageningen, 1986).
de Graaf, N.R., "Natural Regeneration of Tropical Rain Forest in Suriname as a Land-use Option," in *Netherlands Journal of Agricultural Science* 35 (1986): 71–74.

Harvesting the Floodplain Forests
Anderson, A., A. Gely, J. Strudwick, G.L. Sobel and M.G.C. Pinto, "Um Sistema Agroforestal na Varzea do Estuario Amazonico (Ilha das Oncas, Municipio de Barcarena, Estado de Pará)," in *Acta Amazonica* 15 (1985): 195–224.
Anderson, A.B., and A. Gely, "Extractivism and Forest Management by Rural Inhabitants in the Amazon Estuary," in *Natural Resource Management by Indigenous and Old Societies in Amazonia*, ed. D.A. Posey and W. Balee (Bronx, NY: New York Botanical Society, in press).

Extractive Reserves: A Sustainable Development Alternative for Amazonia
Fearnside, P., "Spatial Concentration of Deforestation in the Brazilian Amazon," in *Ambio* 15 (1986): 74–81.
Hecht, S., "Cattle Ranching in Amazonia: Political and Ecological Considerations," in *Frontier Expansion in Amazonia*, ed. M. Schmink and C. Wood (Gainsville: University of Florida Press, 1985).
Schwartzman, S., and M.H. Allegretti, "Extractive Production in the Amazon and the Rubber Tappers' Movement," in *The Social Dynamics of Deforestation: Processes and Alternatives*, ed. S. Hecht and J. Nations.
Schwartzman, S., and M.H. Allegretti, *Extractive Reserves: A Sustainable Development Alternative for Amazonia*, report to the World Wildlife Fund-US, 1987.

Natural Forest Regeneration and Paper Production
Anon, *Ninth Annual Report: Forest Research in the Bajo Calima Concession* (Celulosa y Papel de Colombia SA, Apdo Aereo 6574, Cali, Colombia, 1985).

Sustained Yield Management of Natural Forests in the Palcazu Development Project
Aguilar, D.P.R., "Yanachaga-Chemillen: Futuro Parque Nacional en la Selva Central del Peru," in *Bol. Lima* 45 (1986): 7–21.
Hartshorn, G.S., *La Dinamica de los Bosques Neotropicales*, Tropical Science Center paper no. 8, 1984.
Hartshorn, G.S., R. Simon and J. Tosi Jr, *Sustained Yield Management of Natural Forests: Synopsis of the Palcazu Development Project in the Central Selva of the Peruvian Amazon Region* (1986 unpublished paper available from Tropical Science Center, San Jose, Costa Rica).

The Conservation of Oku Mountain Forests for Wildlife, Watershed, Medicinal Plants and Honey
Macleod, H.L., *Conservation of Oku Mountain Forest* (International Council for Bird Preservation, 1986).

CHAPTER 4: TROPICAL FOREST RESTORATION

Agroforestry and Outreach Project
Carty, W.P., "The Regreening of Haiti: Is Tree-cropping the Answer?" in *Americas* 35 (5) (1983).
Conway, F.J., "Case Study: The Agroforestry Outreach Project in Haiti," in *Conference on Sustainable Development* (London: Earthscan, in press).

Growing Forest from Habitat Fragments in Guanacaste National Forest
Janzen, D.H., *Guanacaste National Park: Tropical, Ecological and Cultural Restoration* (San Jose: Editorial Universidad Estatal Distancia, 1986).

Reafforestation of Amazonian Mines Using Native Species
Knowles, O.H., *Bauxite Mine Restoration in the Amazon*, conference proceedings, National Association of State Land Reclamationists, Orlando, Florida, 1985.
Knowles, O.H., *Preliminary Considerations for the Rehabilitation of the Bauxite Mines at Porto Trombetas, Pará, Brazil*, second annual ecology symposium, Belém, Pará, Brazil, 1978.

Xiaoliang Water and Soil Conservation Project
Tu Men-zhao *et al.*, "Soil Properties in Relation to Changing of Forest Vegetation on Coastal Hilly Land in Guangdong Province," in *Acta Botanica Austro Sinica* 1 (1983): 95–101.

Chen Mao-qian *et al.*, "The Terrestrial Vertebrates in Xiaoliang Tropical Artificial Forest," in *Tropical and Subtropical Forest Ecosystems* 2 (1984): 202–213.

Kuang Lu-ji *et al.*, "The Temperature Humidity Characteristics of the Different Types of Artificial Forests in Xiaoliang, Guangdong," in *Tropical and Subtropical Forest Ecosystems* 2 (1984): 114–121.

Tu Meng-zhao *et al.*, "The Studies on the Influence of Soil pH by Different Plant Communities," in *Tropical and Subtropical Forest Ecosystems* 2 (1984):
110–113.

The Forest Villages of Thailand
Boonkird, S.A., E.C.M. Fernandes and P.K.R. Nair, "Forest Villages: An Agroforestry Approach to Rehabilitating Forest Land Degraded by Shifting Agriculture," in *Agroforestry Systems* 2 (1985): 87–102.

Uganda Forest Village Project
CARE, "Uganda Village Forestry Project Evaluation," (New York: CARE, 1986, unpublished paper).

Struhsacker, T., "Forest Resources and Conservation in Uganda," in *Biological Conservation* (1987): 209–234.

RECOMMENDED READING

Chapter 1

Caufield, C., *In the Rainforest* (New York: Alfred A. Knopf, 1985).

Denslow, J.S., and C. Padoch, *Tropical Forest Peoples* (Berkeley and Los Angeles: University of California Press, 1988).

Ekholm, E., *The Dispossessed of the Earth: Land Reform and Sustainable Development*, Worldwatch paper no. 30, 1979.

Goodland, R., and H. Irwin, *Amazon Jungle: Green Hell to Red Desert?* (New York: Elsevier Scientific, 1975).

Hong, E., *Natives of Sarawak* (Malaysia: Institut Masyrakat, 1987).

IIED/WRI, *World Resources 1987* (New York: Basic Books, 1987).

Myers, N., *The Primary Source: Tropical Forests and our Future* (New York: W.W. Norton and Co, 1984).

Myers, N., *Gaia, an Atlas of Planet Management* (New York: Anchor Books, 1984).

National Academy of Sciences, *Conversion of Tropical Moist Forests* (Washington, DC: NAS, 1980).

National Academy of Sciences, *Ecological Aspects of Development in the Humid Tropics* (Washington, DC: NAS, 1982).

Office of Technology Assessment, *Technologies to Sustain Tropical Forest Resources* (Washington, DC: US Congress, 1984).

Schwartzman, S., *Bankrolling Disasters* (Washington, DC: Sierra Club, 1986).

Stone, R.O., *Dreams of Amazonia* (New York: Viking Penguin, 1985).

Turnbull, C., *The Forest People: A Study of the Pygmies of the Congo* (New York: Simon & Shuster, 1962).

Chapter 2

Cultural Survival Quarterly, special issue on parks and people, 1985.

IUCN, *Buffer Zone Management in Tropical Rainforest Protected Areas* (Gland: IUCN, 1988).

McNeeley, J.A., and K.R. Miller, *National Parks Conservation and Development: The Role of Protected Areas in Sustaining Society* (Washington, DC: Smithsonian Institution Press, 1984).

UNESCO, *Action Plan for Biosphere Reserves* (Paris: UNESCO, 1984).

Chapter 3

Chapin, M., "The Seduction of Models: Chinampa Technology Transfer in Mexico," in *Grassroots Development* No. 1 (1988).

Dover, M., and L.M. Talbot, *To Feed the Earth: Agroecology for Sustainable Development* (Washington, DC: World Resources Institute, 1987).

Janzen, D., "Tropical Agro-ecosystems," in *Science* 181 (1973): 1111–1119.

Lappé, F., and J. Collins, *Food First* (New York: Ballantine Books, 1978).

Nair, P.K.R., *Classification of Agroforestry Systems* (Nairobi: International Council for Research in Agroforestry, 1985).

Chapter 4

Biotropica, special issue on tropical succession, Vol. 12, no. 2 (1980).

Morgan, F., and J.R. Vincent, *Natural Management of Tropical Moist Forests: Silvicultural and Management Prospects of Sustained Utilization* (New Haven: Yale School of Forestry, 1987),

Nectoux, F., and N. Dudley, *A Hardwood Story* (London: Friends of the Earth, 1987).

Sutton, S.L., T.C. Whitmore and A.C. Chadwick, *Tropical Rainforests: Ecology and Management* (Oxford: Blackwell Scientific, 1983).

Whitmore, T.C., *Tropical Rainforests of the Far East* (2nd ed.) (Oxford University Press, 1984).

Wyatt-Smith, J., *et al.*, "Manual of Malayan Silviculture for Inland Forests," in *Malaysian Forestry Record* No. 23 (1963).

INDEX

Academia Sinica, 180, 181–2
Africa, 36, 37, 99. *See also under names of countries*
Agency for International Development, 65
agriculture, large-scale, 41–3, 103
agriculture, sustainable, 102–37
agroforestry, 105, 107, 129–31, 167–70
AID, 159
Amazon:
    alternative resource extraction, 149–52;
    bauxite mines, 173–5;
    deforestation, 37
    See also following entry
Amazon Basin:
    cattle ranching, 28
    colonization, 40
    forest regeneration, 164, 176–7
    indigenous peoples, 48
    Japanese in, 125–7
    rainfall, 58;
    road network, 40–1;
    soils in, 26
Ambuklao, 49
America. *See* United States of America
Amigos de Sian Ka'an, 71–2
Amistad Reserve, La, 62, 63, 78–80
Amuesha people, 157, 159
ANAI, 76, 77
Andes, 32, 37, 43, 84
aquaculture, 109

Asia, Southeast, 30, 31, 32, 37, 41, 45, 91, 139
Audubon societies, 65–6, 75
Awa people, 64, 84, 85
Aztec people, 112

Baboon Sanctuary, 61, 62–5, 67
Bajo Calima, 141, 142, 153–5
bananas, 41
Barro Colorado Island, 58
bauxite mines, 164, 173–5
Belgium, 169, 170
Belize, 32, 36, 66, 112. *See also* Baboon Sanctuary
Belize River, 72
biological diversity, 59
black howler monkeys. *See* Baboon Sanctuary
Bora people, 47, 105, 109, 127–9
Borneo, 47, 48
Brazil:
    debts, 45;
    deforestation, 28, 36;
    forest colonization, 40–1;
    forests, size of, 33;
    land ownership, 41, 42
    See also Amazon Basin; Kayapó people; Tomé Açu
brazil nut gatherers, 149
Brillo Nuevo, 128
Burma, 45

Caboclos, 143, 148–9
Cameroon, 36. *See also* Korup Reserve; Mount Oku

Canada, 33, 169, 170
CARE, 169, 187, 188, 189
Cariari, 116–18
Caribbean, 41, 45
Carton de Colombia SA, 153–5
cattle ranching, 18, 40, 42–3, 113, 115, 116–18, 176
Cauca Valley, 49, 142
Cebu, 166
Celos system, 142
Center for Tropical Agricultural Research and Training (CATIE), 80, 83, 118
China, 25, 28. *See also* Xiaoliong
Colombia, 36, 43–4, 49, 153–5
Colombia–Ecuador Binational Reserve, 62, 63, 83–5
Congo Basin, 26
conservation:
    development and, 14–15, 18; subsidizing,15, 64
Conservation International, 68
Costa Rica, 18, 36, 58, 66, 67, 78, 116–18. *See also* Gandoca/Manzanillo; Guanacaste National Park
crop diversity, 106–7
Cultural Survival, 85
Cuyabeno Reserve, 62, 63, 64, 67, 85–8

debts, 44–5, 65, 67–8
deforestation:
    causes, 11, 37–44;
    effects of, 25–6, 28, 46–51;
    forest types and, 19;
    historical, 25–6;
    indigenous peoples and, 47–8;
    land use after, 32–33;
    law and, 40;
    poverty and, 11, 17;
    ranching and, 104, 116–18;
    rate of, 26, 33–7
Dominican Republic, 42, 167, 169
drip agriculture systems, 106, 114

drought, 49
drugs, 50–1, 57
Dumoga Bone Reserve, 62, 63, 67, 94–6
Dyak people, 47, 48

East–West Center, 98
economic development, 12, 37
Ecuador, 36, 67, 68. *See also* Colombia–Ecuador Binational Reserve; Cuyabeno Reserve; Plan Bosque
education, 62, 63, 72
Efe people, 47
elephants, 91, 92–3
El Salvador, 36, 42
Embera people, 83
EMBRAPA, 177
energy production, 143
erosion: deforestation and, 30, 48–9

fertilizers, 32, 33, 41, 44, 104, 107–8
fires, 45
floods, 49
Food and Agriculture Organization, 69
Ford Foundation, 97, 98, 125, 149
forest management, 25–6:
    age structure, 29
    constraints on, 29–31
    nutrient loss and, 31
forest reserves:
    establishment, 61–3;
    failures of, 60;
    financial support, 64–9;
    function, 57–60;
    multiple-use, 60–1;
    placement, 59–60;
    size, 59–60;
    successes of, 61–5
forests:
    colonizing, 39–41;
    restoration, 163–89;
    sensitivity of, 26–33;

forests—*cont.*
  species, numbers of, 30, 57;
  watersheds and, 58
  See also previous entries and
    deforestation; logging,
    regeneration; soil
Forest villages (Thailand), 164, 167,
  182–4
Foundation for Higher Education,
  85
Friends of the National Zoo, 17
Fundacion Neotropica, 172

Gabon, 38
Gandoca/Manzanillo, 62, 63, 76–7
gardens, 107
Ghana, 36, 38
Grande Carajas, 37
Guanacaste National Park, 164,
  170–2
Guatemala, 66, 112

Haiti, 28, 164, 166, 167–70
Helvetas, 170
home gardens, 132–4
Honduras, 42
hydroelectricity, 41, 156

Iguana ranching, 105, 109, 118–22
Indonesia, 32, 33, 36, 38, 40, 45, 48,
  96–8, 139, 140. *See also*
    Dumoga Bone Reserve
Inter-American Foundation, 83,
  122, 170
irrigation, 58
Ituri forest, 47
Ivory Coast, 36, 38, 49

James Smithson Society, 122
Japan, 38, 125–7. *See also* Tomé
  Açu
Java, 105, 108, 109, 132–4
Jessie Smith Noyes Foundation, 77,
  127

Juanita Valley, 65–6

Kalimantan, 45, 140
Karsted Limestone areas, 113–16
Kayapó people, 47, 105, 106, 108,
  109, 122–5
Khao Yai Reserve, 62, 63, 67, 91–4
Kibale Forest Reserve, 187
Korea, South, 38
Korup Reserve, 62, 63, 99–101
Kuna Wildlands Project, 82
Kuna Yala Comarca, 61, 62, 63–4,
  81–3

land ownership, 40, 41, 48, 63, 71,
  76–7, 103
leaf litter, 114, 115–16
leafspot, 51
logging:
  effects of, 48, 139, 144;
  extent of, 38;
  non-destructive, 18, 153–5;
  reasons for, 37–9;
  regeneration and, 27–9;
  selective, 30, 33, 37;
  species numbers and, 30;
  timber prices, 38
Lynch, James, 113

MacArthur Foundation, 66, 83, 172
Madagascar, 37
mahogany, 30, 89
Malaysia, 36, 38, 45, 58–9, 139
Malaysian Uniform Systems, 31,
  140, 141
Manaus, 59
Manhattan Project, 16
Manu Reserve, 67, 88–90
Maya people, 47, 105, 108, 109,
  110–12, 114, 115, 125
Mayhew, Henry, 11–12
Mbuti, 47

Mexico, 47, 106, 112. *See also* Karsted Limestone areas; Sian Ka'an Reserve
Middle East, 25
milpas, 111–12, 113, 114
Mineraçao Rio do Norte, 173, 174, 175
mining, 37–8, 173–5
monocultures, 26, 32, 103, 106
Mt Cyclops Reserve, 62, 63, 97
Mt Oku, 67, 142, 143, 160–2

National Geographic Society, 125, 177
National Institutes of Health, 101
National Science Foundation, 125, 127, 177
National Zoological Park, 14
Natural Resources Institute, 122
Nature Conservancy, 66, 72, 157, 172
New Guinea, 28, 105, 106, 108, 134–7
Nigeria, 36, 38

oil prices, 40, 45
Organization for Tropical Studies, 58, 66
Ormu Besar, 97, 98
Overseas Development Administration, 101

PACT, 170
Palcazu project, 141, 157–60
Palmira, 156
Panama, 36, 78, 109. *See also following entry and* Kuna Yala Comarca
Panama Canal, 27–8, 49, 58
Pan American Development Foundation, 169, 170
Papua New Guinea, 36
Paragominas, 176

Pasoh Forestry Research Center, 58–9
Pax World Foundation, 170
Peace Corps Volunteers, 66
Peru, 41–2, 66, 105, 129–31. *See also* Bora people; Manu Reserve; Palcazu project
pesticides, 33, 41, 44, 104, 108
Pew Charitable Trust, 172
pharmaceutical industry, 50–1
Philippines, 38, 49, 67, 166; soil conservation, 164, 184–7
Plan Bosque, 164, 177–9
population, 44
Port au Prince, 168
poverty, 11, 43
Pulpapel, 142
pygmies, 47

ranching, 103–4, 150. *See also* cattle ranching
regeneration: aiding, 163–7; difficulties of, 27–9; time needed for, 39
research, 14, 15, 16, 58, 63, 144, 164
resettlement programmes, 40
Rhodes, R., 16
rice, 32, 41, 43–4, 63, 95
Rio Beni Reserve, 68
Rio Nima watershed, 143, 156
Romans, 25
Royal Entomological Society, 96
rubber, 31, 41, 143, 149, 150, 152
Rubinoff, I., 15, 69

Sabah, 140, 141
Sahel, 25
Santa Rosa National Park, 172
Sarawak, 140, 141
Secoya people, 85

seeds:
  banks, 29, 58;
  spreading, 28
Selva, La, 58
Shell, 170
shifting agriculture, 102–3, 134–7.
    See also slash and burn;
    swidden-fallow agriculture
Sian Ka'an Reserve, 62, 63, 67,
    70–2, 115
Siona people, 85, 87
Sister Sponsorship, 65–6
slash and burn, 39, 102–3, 114,
    118, 124, 135, 160
Smithsonian Institution, 14, 17, 58,
    69, 83, 113, 122
soil:
  classification, 123;
  compaction, 30, 139;
  conservation, 164, 184–7;
  deforestation and, 27, 30, 31, 32;
  erosion, 30, 48–9, 58, 135, 139;
  nutrients and, 26–7, 31, 32
species: destruction of, 13, 49–51
species-banking, 15
successional schemes, 106
Suriname, 36, 142, 144–9
Sweden, 172
swidden-fallow agriculture, 102–3,
    105–6, 127–9, 158

Taiwan, 38
Tamshyiyacu, 130
taungya approch, 182–3
Thailand, 36, 38, 67. See also Forest
    villages; Khao Yai Reserve
timber prices, 38
Tomé Açu, 105, 108, 110, 125–7
tourism, 62, 63, 64, 66–7, 71, 74,
    75, 80, 87, 89, 90, 93
triage, 14, 15
Trombetas, 173
Tropical Agricultural Research and
    Training Center, 83

Uganda Village Forests, 164, 166,
    187–9
unemployment, 42
UNESCO Man and Biosphere
    program, 60–1, 89, 129, 131
Union of Soviet Socialist Republics,
    33
United States of America, 25, 26,
    29–30, 33, 38, 51, 68, 138
USAID, 83, 93, 160, 162, 170, 187,
    189

Vietnam War, 45

Wallace, Operation, 96
W. Alton Jones Foundation, 122,
    172
war, 45
watershed conservation, 155–7
West Indies, 37
wildfires, 45
World Bank, 94, 95, 96
World Neighbors, 166, 185, 187
World Resources Institute, 69
World Vision, 170
World Wide Fund for Nature, 101
World Wildlife Fund, 68, 90, 91,
    93, 95, 96, 116, 125, 162
World Wildlife Fund-US, 17, 59,
    66, 72, 75, 77, 83, 85, 88, 97,
    98, 125, 149, 152

Xiaoliang, 164, 166, 180–2

Yanesha Forestry Cooperative, 159

Zaire, 33, 36, 41, 47
Zona Protectora, La, 66

Also Available from Island Press

Land and Resource Planning in the National Forests
By Charles F. Wilkinson and H. Michael Anderson
Foreword by Arnold W. Bolle

This comprehensive, in-depth review and analysis of planning, policy, and law in the National Forest System is the standard reference source on the National Forest Management Act of 1976 (NFMA). This clearly written, nontechnical book offers an insightful analysis of the Fifty Year Plans and how to participate in and influence them.

1987. xii, 396 pp., index.
Paper ISBN 0-933280-38-6. **$19.95**

Reforming the Forest Service
By Randal O'Toole

Reforming the Forest Service contributes a completely new view to the current debate on the management of our national forests. O'Toole argues that poor management is an institutional problem; he shows that economic inefficiencies and environmental degradation are the inevitable result of the well-intentioned but poorly designed laws that govern the Forest Service. This book proposes sweeping reforms in the structure of the agency and new budgetary incentives as the best way to improve management.

1988. xii, 256 pp., graphs, tables, notes.
Cloth, ISBN 0-933280-49-1. **$19.95**
Paper, ISBN 0-933280-45-9. **$34.95**

Last Stand of the Red Spruce
By Robert A. Mello
Published in cooperation with Natural Resources Defense Council

Acid rain—the debates rage between those who believe that the cause of the problem is clear and identifiable and those who believe that the evidence is inconclusive. In *Last Stand of the Red Spruce*, Robert A. Mello has written an ecological detective story that unravels this confusion and explains how air pollution is killing our nation's forests. Writing for a lay audience, the author traces the efforts of scientists trying to solve the mystery of the dying red spruce trees on Camels Hump in Vermont. Mello clearly and succinctly presents both sides of an issue on which even the scientific community is split and concludes that the scientific evidence uncovered on Camels Hump elevates the issues of air pollution and acid rain to new levels of national significance.

1987. xx, 156 pp., illus., references, bibliography.
Paper, ISBN 0-933280-37-8. **$14.95**

Western Water Made Simple, by the editors of **High Country News**
Edited by Ed Marston

Winner of the 1986 George Polk Award for environmental reporting, these four special issues of *High Country News* are here available for the first time in book form. Much has been written about the water crisis in the West, yet the issue remains confusing and difficult to understand. *Western Water Made Simple,* by the editors of *High Country News,* lays out in clear language the complex issues of Western water. This survey of the West's three great rivers—the Colorado, the Columbia, and the Missouri—includes material that reaches to the heart of the West—its ways of life, its politics, and its aspirations. *Western Water Made Simple* approaches these three river basins in terms of overarching themes combined with case studies—the Columbia in an age of reform, the Colorado in the midst of a fight for control, and the Missouri in search of its destiny.

1987. 224 pp., maps, photographs, bibliography, index.
Paper, ISBN 0-933280-39-4. **$15.95**

**The Report of the President's Commission on Americans Outdoors:
The Legacy, The Challenge**
With Case Studies
Preface by William K. Reilly

"If there is an example of pulling victory from the jaws of disaster, this report is it. The Commission did more than anyone expected, especially the administration. It gave Americans something serious to think about if we are to begin saving our natural resources."
—Paul C. Pritchard, President, National Parks and Conservation Association.

This report is the first comprehensive attempt to examine the impact of a changing American society and its recreation habits since the work of the Outdoor Recreation Resource Review Commission, chaired by Laurance Rockefeller in 1962. The President's Commission took more than two years to complete its study; the Report contains over sixty recommendations, such as the preservation of a nationwide network of "greenways" for recreational purposes and the establishment of an annual $1 billion trust fund to finance the protection and preservation of our recreational resources. The Island Press edition provides the full text of the report, much of the additional material compiled by the Commission, and twelve selected case studies.

1987. xvi, 426 pp., illus., appendixes, case studies.
Paper, ISBN 0-933280-36-X. **$24.95**

Public Opinion Polling: A Handbook for Public Interest and Citizen Advocacy Groups
By Celinda C. Lake, with Pat Callbeck Harper

"Lake has taken the complex science of polling and written a very usable 'how-to' book. I would recommend this book to both candidates and organizations interested in professional, low-budget, in-house polling." — Stephanie Solien, Executive Director, Women's Campaign Fund.

Public Opinion Polling is the first book to provide practical information on planning, conducting, and analyzing public opinion polls as well as guidelines for interpreting polls conducted by others. It is a book for anyone—candidates, state and local officials, community organizations, church groups, labor organizations, public policy research centers, and coalitions focusing on specific economic issues—interested in measuring public opinion.

1987. x, 166 pp., bibliography, appendix, index.
Paper, ISBN 0-933280-32-7. **$19.95**
Companion software now available.

Green Fields Forever: The Conservation Tillage Revolution in America
By Charles E. Little

"*Green Fields Forever* is a fascinating and lively account of one of

211

the most important technological developments in American agriculture. . . . Be prepared to enjoy an exceptionally well-told tale, full of stubborn inventors, forgotten pioneers, enterprising farmers—and no small amount of controversy."—Ken Cook, World Wildlife Fund and The Conservation Foundation.

Here is the book that will change the way Americans think about agriculture. It is the story of "conservation tillage"—a new way to grow food that, for the first time, works *with,* rather than against, the soil. Farmers who are revolutionizing the course of American agriculture explain here how conservation tillage works. Some environmentalists think there are problems with the methods, however; author Charles E. Little demonstrates that on this issue both sides have a case, and the jury is still out.

1987. 189 pp., illus., appendixes, index, bibliography.
Cloth, ISBN 0-933280-35-1. **$24.95**
Paper, ISBN 0-933280-34-3. **$14.95**

Federal Lands: A Guide to Planning, Management, and State Revenues
By Sally K. Fairfax and Carolyn E. Yale

"An invaluable tool for state land managers. Here, in summary, is everything that one needs to know about federal resource management policies."—Rowena Rogers, President, Colorado State Board of Land Commissioners.

Federal Lands is the first book to introduce and analyze in one accessible volume the diverse programs for developing resources on federal lands. Offshore and onshore oil and gas leasing, coal and geothermal leasing, timber sales, grazing permits, and all other programs that share receipts and revenues with states and localities are considered in the context of their common historical evolution as well as in the specific context of current issues and policy debates.

1987. xx, 252 pp., charts, maps, bibliography, index.
Paper, ISBN 0-933280-33-5. **$24.95**

Hazardous Waste Management: Reducing the Risk
By Benjamin A. Goldman, James A. Hulme, and Cameron Johnson for the Council on Economic Priorities

Hazardous Waste Management: Reducing the Risk is a comprehensive sourcebook of facts and strategies that provides the analytic tools needed by policy makers, regulating agencies, hazardous waste generators, and host communities to compare facilities on the basis of site, management, and technology. The Council on Economic Priorities' innovative ranking system applies to real-world, site-specific evaluations, establishes a consistent protocol for multiple applications, assesses relative benefits and risks, and evaluates and ranks ten active facilities and eight leading commercial management corporations.

1986. xx, 316 pp., notes, tables, glossary, index.
Cloth, ISBN 0-933280-30-0. **$64.95**
Paper, ISBN 0-933280-31-9. **$34.95**

An Environmental Agenda for the Future
By Leaders of America's Foremost Environmental Organizations

". . . a substantive book addressing the most serious questions about the future of our resources."—John Chafee, U.S. Senator, Environmental and Public Works Committee. "While I am not in agreement with many of the positions the authors take, I believe this book can be the basis for constructive dialogue with industry representatives seeking solutions to environmental problems."—Louis Fernandez, Chairman of the Board, Monsanto Corporation.

The chief executive officers of ten major environmental and conservation organizations launched a joint venture to examine goals that the environmental movement should pursue now and into the twenty-first century. This book presents policy recommendations for implementing the changes needed to bring about a healthier, safer world. Topics discussed include nuclear issues, human population growth, energy strategies, toxic waste and pollution control, and urban environments.

1985. viii, 155 pp., bibliogrpahy.
Paper, ISBN 0-933280-29-7. **$9.95**

Water in the West
By Western Network

Water in the West is an essential reference tool for water managers, public officials, farmers, attorneys, industry officials, and students and professors attempting to understand the competing pressures on our most important natural resource: water. Here is an in-depth analysis of the effects of energy development, Indian rights, and urban growth on other water users.

1985. *Vol. III: Western Water Flows to the Cities*
v, 217 pp., maps, table of cases, documents, bibliography, index.
Paper, ISBN 0-933280-28-9.   **$25.00**

These titles are available directly from Island Press, Box 7, Covelo, CA 95428. Please enclose $2.75 shipping and handling for the first book and $1.25 for each additional book. California and Washington, DC residents add 6% sales tax. A catalog of current and forthcoming titles is available free of charge. Prices subject to change without notice.